AF389359

International Conference on "Holography Meets Advanced Manufacturing"

International Conference on "Holography Meets Advanced Manufacturing"

Editors

Vijayakumar Anand
Amudhavel Jayavel
Viktor Palm
Shivasubramanian Gopinath
Andrei Bleahu
Aravind Simon John Francis Rajeswary
Kaupo Kukli
Vinoth Balasubramani
Daniel Smith
Soon Hock Ng
Saulius Juodkazis

Basel • Beijing • Wuhan • Barcelona • Belgrade • Novi Sad • Cluj • Manchester

Editors

Vijayakumar Anand
University of Tartu
Tartu, Estonia

Amudhavel Jayavel
University of Tartu
Tartu, Estonia

Viktor Palm
University of Tartu
Tartu, Estonia

Shivasubramanian Gopinath
University of Tartu
Tartu, Estonia

Andrei Bleahu
University of Tartu
Tartu, Estonia

Aravind Simon John Francis
Rajeswary
University of Tartu
Tartu, Estonia

Kaupo Kukli
University of Tartu
Tartu, Estonia

Vinoth Balasubramani
King Abdullah University of
Science and Technology
(KAUST)
Thuwal, Saudi Arabia

Daniel Smith
Swinburne University of
Technology
Melbourne, Australia

Soon Hock Ng
Swinburne University of
Technology
Melbourne, Australia

Saulius Juodkazis
Swinburne University of
Technology
Melbourne, Australia

Editorial Office
MDPI
St. Alban-Anlage 66
4052 Basel, Switzerland

This is a reprint of articles from the Proceedings published online in the open access journal *Engineering Proceedings* (ISSN 2673-4591) (available at: https://www.mdpi.com/2673-4591/34/1).

For citation purposes, cite each article independently as indicated on the article page online and as indicated below:

Lastname, A.A.; Lastname, B.B. Article Title. *Journal Name* **Year**, *Volume Number*, Page Range.

ISBN 978-3-0365-8452-2 (Hbk)
ISBN 978-3-0365-8453-9 (PDF)
doi.org/10.3390/books978-3-0365-8453-9

Contents

Preface

The CIPHR group, Institute of Physics, University of Tartu, Estonia, and the Optical Sciences Center, Swinburne University of Technology, Australia, jointly organized the interdisciplinary online conference "Holography meets Advanced Manufacturing", held from 20 to 22 February 2023.

To develop imaging concepts that go beyond the state of the art, there has been significant collaboration in the recent years between the research fields of holography and advanced manufacturing. This was the motivation for organizing the joint international conference with Swinburne University of Technology, a world leader in advanced manufacturing technology.

Vijayakumar Anand, Amudhavel Jayavel, Viktor Palm, Shivasubramanian Gopinath, Andrei Bleahu, Aravind Simon John Francis Rajeswary, Kaupo Kukli, Vinoth Balasubramani, Daniel Smith, Soon Hock Ng, and Saulius Juodkazis

Editors

engineering proceedings

MDPI

Editorial

Statement of Peer Review †

Vijayakumar Anand [1,2,*], Amudhavel Jayavel [1], Viktor Palm [1], Shivasubramanian Gopinath [1], Andrei Bleahu [1], Aravind Simon John Francis Rajeswary [1], Kaupo Kukli [1], Vinoth Balasubramani [3], Ravi Kumar [4], Daniel Smith [2], Soon Hock Ng [2] and Saulius Juodkazis [2,5]

[1] Institute of Physics, University of Tartu, W. Ostwaldi 1, 50411 Tartu, Estonia; amudhavel.jayavel@ut.ee (A.J.); viktor.palm.001@ut.ee (V.P.); shivasubramanian.gopinath@ut.ee (S.G.); andrei-ioan.bleahu@ut.ee (A.B.); aravind@ut.ee (A.S.J.F.R.); kaupo.kukli@ut.ee (K.K.)

[2] Optical Sciences Center, Swinburne University of Technology, Melbourne 3122, Australia; danielsmith@swin.edu.au (D.S.); soonhockng@swin.edu.au (S.H.N.); sjuodkazis@swin.edu.au (S.J.)

[3] Division of Biological and Environmental Sciences and Engineering, King Abdullah University of Science and Technology (KAUST), Thuwal 23955-6900, Saudi Arabia; vinoth.balasubramani@kaust.edu.sa

[4] Department of Physics, SRM University-AP, Amaravati 522502, India; ravi.k@srmap.edu.in

[5] WRH Program International Research Frontiers Initiative (IRFI), Tokyo Institute of Technology, Nagatsuta-cho, Midori-ku, Yokohama 226-8503, Japan

* Correspondence: vijayakumar.anand@ut.ee

† All the papers published in the volume are presented at the International Conference on "Holography Meets Advanced Manufacturing", Online, 20–22 February 2023.

Citation: Anand, V.; Jayavel, A.; Palm, V.; Gopinath, S.; Bleahu, A.; John Francis Rajeswary, A.S.; Kukli, K.; Balasubramani, V.; Kumar, R.; Smith, D.; et al. Statement of Peer Review. *Eng. Proc.* **2023**, *34*, 28. https://doi.org/10.3390/engproc2023034028

Published: 25 July 2023

In submitting conference proceedings to *Engineering Proceedings*, the editors of the proceedings certify to the publisher that all papers published in this volume have been subjected to peer review. Reviews were conducted by expert referees and met the professional and scientific standards expected of a proceedings journal.

- Type of peer review: single-blind.
- Conference submission management system: A total of 54 abstracts were submitted for the conference, and 27 papers were published in *Engineering Proceedings* (https://sciforum.net/dashboard/event_chair/submissions/HMAM2).

All of the submissions were processed completely online using MDPI's SciForum platform. This includes the submission of abstracts, presentation files, and manuscripts; a preliminary check of the format and template by event organizers; an initial peer review of the submissions by event chairs; a full peer review by external international reviewers; the submission of revised version by the authors; and the final check and acceptance by the reviewers and event chairs. Further checks for compliance to MDPI standards involving plagiarism were carried out by the editors of the *Engineering Proceedings* journal of MDPI.

Number of submissions sent for review: 35;

Number of submissions accepted: 27;

Acceptance rate (number of submissions accepted/number of submissions received): 0.77;

Average number of reviews per paper: 2;

Total number of reviewers involved: 24 reviewers.

Here is some additional information on the review process:

1. Participants received the results of the review. Types of decisions: reject, resubmit, revision, and acceptance:

Reject: Following peer review, the paper was judged not acceptable for the Holography Meets Advanced Manufacturing (HMAM2) conference;

Resubmit: The submitted version of the paper was not acceptable and required major revisions, but there was clear potential in the paper, and HMAM2 was prepared to consider a new version. Authors were offered an opportunity to revise their manuscripts and resubmit as a new submission. The previous reviewer comments and editorial recommendations were applied, and then in the next round the editors and reviewers were convinced by the authors' responses and revisions that their paper fit the standards of HMAM2.

Revision: The paper required changes before a final decision could be made. Authors were suggested to revise their manuscripts in light of comments received from reviewers and editors, and advised to submit a revised version with suitable responses to reviewer comments for consideration. Revisions underwent a further peer review process in some cases, and in some other cases, the submissions underwent another round of revision.

Final Acceptance: The paper was found acceptable for publication when all of the required standards of HMAM2 were met. Some time before final acceptance, sub-editing changes and minor amendments were suggested to ensure the paper fully matched our criteria. After final checking by the Editorial Office, acceptance was confirmed, and the paper was forwarded to the publishers for publication.

2. The accepted and presented submissions were published in MDPI's *Engineering Proceedings*. Selected submissions were invited to the Special Issue on "Research in Computational Optics", which was organized by Prof. Vijayakumar Anand, Prof. Ravi Kumar, Dr. Andra Naresh Kumar Reddy, and Dr. Vinoth Balasubramani and was published in MDPI's *Photonics* (IF—2.536) via APC waivers and special discounts (https://www.mdpi.com/journal/photonics/special_issues/B3GWFOP55V). In the Special Issue, three extended versions of the conference submissions have already been published.

Funding: This conference was funded by European Union's Horizon 2020 research and innovation programme grant agreement No. 857627 (CIPHR).

Conflicts of Interest: The authors declare no conflict of interest.

engineering proceedings

Editorial

Preface: International Conference on Holography Meets Advanced Manufacturing (HMAM2) [†]

Vijayakumar Anand [1,2,*], Amudhavel Jayavel [1], Viktor Palm [1], Shivasubramanian Gopinath [1], Andrei Bleahu [1], Aravind Simon John Francis Rajeswary [1], Kaupo Kukli [1], Vinoth Balasubramani [3], Daniel Smith [2], Soon Hock Ng [2] and Saulius Juodkazis [2,4]

[1] Institute of Physics, University of Tartu, 50411 Tartu, Estonia; amudhavel.jayavel@ut.ee (A.J.); viktor.palm.001@ut.ee (V.P.); shivasubramanian.gopinath@ut.ee (S.G.); andrei-ioan.bleahu@ut.ee (A.B.); aravind@ut.ee (A.S.J.F.R.); kaupo.kukli@ut.ee (K.K.)

[2] Optical Sciences Center, Swinburne University of Technology, Melbourne 3122, Australia; danielsmith@swin.edu.au (D.S.); soonhockng@swin.edu.au (S.H.N.); sjuodkazis@swin.edu.au (S.J.)

[3] Division of Biological and Environmental Sciences and Engineering, King Abdullah University of Science and Technology (KAUST), Thuwal 23955-6900, Saudi Arabia; vinoth.balasubramani@kaust.edu.sa

[4] Tokyo Tech World Research Hub Initiative (WRHI), School of Materials and Chemical Technology, Tokyo Institute of Technology, Tokyo 152-8550, Japan

[*] Correspondence: vijayakumar.anand@ut.ee

[†] All the papers published in the volume are presented at the International Conference on "Holography Meets Advanced Manufacturing", Online, 20–22 February 2023.

Abstract: The CIPHR group, Institute of Physics, University of Tartu, Estonia, and Optical Sciences Center, Swinburne University of Technology, Australia, jointly organized the interdisciplinary online conference "Holography Meets Advanced Manufacturing" during 20–22 February 2023.

Keywords: digital holography; incoherent holography; computational imaging; microscopy; diffractive optics; quantitative phase imaging; structured light; tomography; optical security; cryptography; laser beam shaping; metalenses; micro/nanofabrication; femtosecond fabrication; OAM beams; optoelectronic materials and devices; non-diffracting beams; space–time correlations; non-linear optics

Citation: Anand, V.; Jayavel, A.; Palm, V.; Gopinath, S.; Bleahu, A.; John Francis Rajeswary, A.S.; Kukli, K.; Balasubramani, V.; Smith, D.; Ng, S.H.; et al. Preface: International Conference on Holography Meets Advanced Manufacturing (HMAM2). *Eng. Proc.* **2023**, *34*, 29. https://doi.org/10.3390/engproc2023034029

Published: 24 July 2023

1. Introduction

To develop imaging concepts that go beyond the state-of-the-art, there has been significant collaboration in recent years between the research fields of holography and advanced manufacturing. This was the motivation for organizing the joint international conference between CIPHR group, Institute of Physics, University of Tartu and Optical Sciences Center, Swinburne University of Technology, a world leader in advanced manufacturing technology. The first day of the conference was dedicated to holography; the second day focused on advanced manufacturing; and the third day was devoted to industry applications.

This conference provided a platform for exchanging ideas, interdisciplinary, international collaboration, with three keynote talks from three world-renowned researchers, several invited talks from leading researchers across the globe, poster presentations of students, and industry-ready ideas for industry linkages. The submissions were accepted from the broad areas of holography and advanced manufacturing, which included digital holography, incoherent holography, computational imaging, microscopy, diffractive optics, quantitative phase imaging, structured light, tomography, optical security, cryptography, laser beam shaping, metalenses, micro/nanofabrication, femtosecond fabrication, OAM beams, optoelectronic materials and devices, non-diffracting beams, and space–time correlations and non-linear optics.

The keynote talks were scheduled for 35 min each and 10 min for a question session. The invited talks were scheduled for 15 min each, followed by 5 min for questions, while the poster presentations were allotted a duration of 10 min each, followed by 2 min for questions. Each day of the conference was divided into three sessions.

All the submissions were processed completely online using the Sciforum platform operated by MDPI. This includes the submission of abstracts, presentation files, and manuscripts, preliminary checks of format and template by event organizers, initial peer review of the submissions by event chairs, full peer review by external international reviewers, submissions of revised versions by the authors, and final checks and acceptance by the reviewers and event chairs. Further checks of compliance to MDPI standards including plagiarism checks were performed by the editors of *Engineering Proceedings*, published by MDPI. All accepted and presented submissions were published in *Engineering Proceedings*. Selected submissions were invited to the Special Issue on "Research in Computational Optics", organized by Prof. Vijayakumar Anand, Prof. Ravi Kumar, Dr. Andra Naresh Kumar Reddy, and Dr. Vinoth Balasubramani in the MDPI journal *Photonics* (IF–2.536), with APC waivers and special discounts https://www.mdpi.com/journal/photonics/special_issues/B3GWFOP55V (accessed on 21 July 2023). In the Special Issue, three extended versions of the conference submissions have already been published.

2. Conference Organizing Committee

The following committees namely event organizers, event committee and event chairs for the three days are shown in Tables 1–3 respectively. were formed in October 2022 to effectively coordinate the three-day conference, and the event committee frequently met to review the procedures and steps needed to successfully carry out three-day conference.

Table 1. Event Organizers.

Sl. No	Name	Position and Affiliations
1.	Aravind Simon J R	Research Manager, CIPHR Group, Institute of Physics, University of Tartu, Estonia
2.	Tiia Lillemaa	Project Manager CIPHR Group, Institute of Physics, University of Tartu, Estonia

Table 2. Event Committee.

Sl. No	Name	Position and Affiliations
1.	Prof. Vijayakumar Anand	Associate Professor and ERA Chair Holder, CIPHR Group, Institute of Physics, University of Tartu, Estonia University of Tartu
2.	Dr. Amudhavel Jayavel	Researcher, CIPHR Group, Institute of Physics, University of Tartu, Estonia
3.	Dr. Viktor Palm	Researcher, CIPHR Group, Institute of Physics, University of Tartu, Estonia
4.	Aravind Simon J R	Research Manager, CIPHR Group, Institute of Physics, University of Tartu, Estonia
5.	Tiia Lillemaa	Project Manager, CIPHR Group, Institute of Physics, University of Tartu, Estonia
6.	Andrei-Ioan Bleahu	Junior researcher, CIPHR Group, Institute of Physics, University of Tartu, Estonia
7.	Shivasubramanian Gopinath	Junior Researcher, CIPHR Group, Institute of Physics, University of Tartu, Estonia
8.	Daniel Smith	Junior Researcher, Optical Sciences Center, Swinburne University of Technology, Australia

Table 3. Event chairs.

Sl. No	Name	Affiliations
		DAY 1
1.	Prof. Vijayakumar Anand	Associate Professor and ERA Chair Holder, CIPHR Group, Institute of Physics, University of Tartu, Estonia
2.	Dr. Vinoth Balasubramani	Senior researcher at King Abdullah University of Science and Technology (KAUST), Saudi Arabia
		DAY 2
1.	Dr. Soon Hock Ng	Senior researcher at Swinburne University of Technology, Australia
2.	Prof. Kaupo Kukli	Laboratory of Thin Film Technology, Institute of Physics, University of Tartu, Estonia
		DAY 3
1.	Dr. Amudhavel Jayavel	Researcher, CIPHR Group, Institute of Physics, University of Tartu, Estonia
2.	Dr. Viktor Palm	Researcher, CIPHR Group, Institute of Physics, University of Tartu, Estonia

3. Invitations and Participants

Approximately 30 leading researchers specializing in computational imaging from around the world received invitations to deliver an invited talk on the first day of the conference. We invited 25 leading researchers in advanced manufacturing to present an invited talk on day two of the conference. Approximately 50 industry members, mostly medical doctors and surgeons, were invited to give presentations on industry applications and their requirements. Several medical doctors and surgeons shared their difficulties in utilizing imaging and characterization tools, as well as features and capabilities they hope to see in the medical equipment and tools in the future. We sent out invitations to attend the conference and give poster presentations to more than 100 universities worldwide, more than 50 companies in Estonia and more than 100 companies globally.

More than 140 participants, including event chairs and organizers (11), keynote speakers (3), opening speakers (3), invited presenters (29), poster presenters (23), and registered participants (95) from different universities, companies, researchers, research scholars, and master's students from across the globe participated in the conference.

4. Conference Programme

Day 1 of the conference was dedicated to Holography.

Prof. Vijayakumar Anand started Day 1 of the conference by introducing the online conference on "Holography meets advanced manufacturing". Prof. Toomas Plank, Director of the Institute of Physics, University of Tartu, gave a warm welcome before the opening talk by Prof. Peeter Saari, Professor Emeritus, University of Tartu, on the topic of "Optics in Estonia: Research and Innovation Highlights". Prof. Joseph Rosen from the School of Electrical and Computer Engineering, Ben-Gurion University of the Negev, Israel, delivered the keynote talk on "Advanced Imaging Methods Using Coded Aperture Digital Holography". Three sessions were held. Session 1 had six invited talks, presided over by Session Chair Prof. Vijayakumar Anand and Co-Chair Daniel Smith. Session 2 also featured six invited talks, and was presided over by Session Chair Dr. Vinoth Balasubramani and Co-Chair Andrei Bleahu. Prof. Vijayakumar Anand and Dr. Vinoth Balasubramani presided over Session 3 of the poster presentations. In the poster sessions, eight junior researchers presented their research. In all sessions, the presenters were introduced by the Co-Chairs.

Day 2 of the conference

Prof. Vijayakumar Anand started Day 2 of the conference by giving a brief introduction to the program schedule. Prof. Saulius Juodkazis, Deputy Director of the Optical Sciences Center and Director of Nanotechnology, Swinburne University of Technology, Melbourne, Australia, presented the keynote talk on "Ultra-short laser pulses as material synthesis and lithography tool". Three sessions were held on the second day. Session 1 had five invited talks and was presided over by Session Chair Prof. Vijayakumar Anand and Co-Chair Daniel Smith. Session 2 featured four invited talks and was presided over by Chair Prof. Kaupo Kukli and Co-Chair Andrei Bleahu. Prof. Vijayakumar Anand and Prof. Kaupo Kukli presided over Session 3 of the poster presentations. In the poster sessions, six research scholars presented their work. In all sessions, the presenters were introduced by the Co-Chairs.

Day 3 of the conference

Dr. Amudhavel Jayavel started Day 3 of the conference by introducing the program schedule. Prof. Heli Valtna presented the keynote talk on "Mapping the path from idea to economically scalable application in 3D imaging and sensing". Three sessions were held on the third day. Session 1 had four invited talks from medical doctors and was presided over by Chair Dr. Amudhavel Jayavel, and Co-Chair Andrei Bleahu. Session 2 featured four invited talks and was presided over by Chair Amudhavel Jayavel and Co-Chair Shivasubramanian Gopinath. Dr. Amudhavel Jayavel and Dr. Viktor Palm presided over Session 3 of the poster presentation. In the poster sessions, eight research scholars presented their work. All presentations are available on the YouTube channel CIPHR Talkies, and are permanently stored on the MDPI Sciforum platform.

The following shown in Table 4 is a complete conference schedule, including each speaker's name, the topic they presented, and their affiliations.

Table 4. Complete conference schedule.

Day 1: Monday, 20 February 2023 HOLOGRAPHY		
09:00–09:15	Virtual conference window open	
09:15–09:30	Introduction	Prof. Vijayakumar Anand Associate Professor, University of Tartu
09:30–09:40	Welcome note	Prof. Toomas Plank Director, Institute of Physics, University of Tartu
09:40–10:05	Opening talk: "Optics in Estonia: Research and innovation highlights"	Prof. Peeter Saari Professor Emeritus, University of Tartu, Estonia
10:10–10:55	Keynote: "Advanced imaging methods using Coded aperture digital holography"	Prof. Joseph Rosen School of Electrical and Computer Engineering, Ben-Gurion University of the Negev, Israel
Session 1—H1. Holography Chairs: Prof. Vijayakumar Anand and Daniel Smith		
11:00–11:20	Incoherent digital holography for multidimensional motion-picture imaging	Dr. Tatsuki Tahara Senior Researcher, National Institute of Information and Communications Technology (NICT), Japan
11:25–11:45	Synchrotron-FTIR Technique to underpin the future of advanced manufacturing technology and its applications at Australian Synchrotron	Dr. Jitraporn (Pimm) Vongsvivut Senior Beamline Scientist, Australian Synchrotron, Victoria, Australia
11:50–12:10	3D and see-through light-field display using digitally printed holographic screen	Prof. Jackin Boaz Jessie Associate Professor, Kyoto Institute of Technology, Kyoto, Japan
12:15–12:35	Manipulating light with micro and nano-optics and holographic techniques	Prof. Shanti Bhattacharya Professor, Department of Electrical Engineering, IIT Madras, India
12:40–13:00	Digital polarization holography: challenges and opportunities	Prof. Rakesh Kumar Singh Associate Professor, Department of Physics, IIT (BHU), India
13:05–13:25	Information security: an optical approach	Prof. Ravi Kumar Assistant Professor, Department of Physics, SRM University-AP, Andhra Pradesh, India
Session 2—H2. Holography Chairs: Dr. Vinoth Balasubramani and Andrei Bleahu		
13:30–13:50	Role of deep learning in optical imaging	Prof. Inbarasan Muniraj Head of LiFE Lab, Alliance University, Bengaluru, India
13:55–14:15	Sapphire diffractive axicon milled with femtosecond laser ablation for imaging applications	Daniel Smith Poster Presentation Optical Sciences Center, Swinburne University of Technology, Australia
14:20–14:40	Lensless hyperspectral phase retrieval via alternating direction method of multipliers and spectral proximity operators	Dr. Igor Shevkunov Postdoctoral researcher, Tampere University, Finland
14:45–15:05	Quantitative analysis of illumination and detection corrections in adaptive light sheet fluorescence microscopy	Dr. Mani Ratnam Rai Postdoctoral Researcher, NC State University/UNC-Chapel Hill, Raleigh, USA
15:10–15:30	Depth resolved imaging by digital holography via sample-shifting	Prof. Suhas Veetil Associate Professor, School of Engineering Science and Technology at Higher Colleges of Technology, UAE
17:30–17:50	Spin-orbit modal shaping of vortex beams	Prof. Etienne Brasselet Research director, CNRS, Laboratoire Ondes et Matière d'Aquitaine, University of Bordeaux, France
17:55–18:15	Light sheet fluorescence microscopy using incoherent light detection	Prof. Mariana Potcoava Research Assistant Professor, University of Illinois at Chicago, USA

Table 4. *Cont.*

	Poster Presentations Session Chairs: Prof. Vijayakumar Anand and Dr. Vinoth Balasubramani	
	15:45–18:00 (10 min for presentation + 2 min for Questions)	
15:45–16:00	Single shot lensless interferenceless phase imaging of biochemical samples using Synchrotron near infrared beam	Molong Han Optical Sciences Center, Swinburne University of Technology, Australia
16:00–16:15	Digital Fourier transform holography using a beam displacer	Mohit Rathor Laboratory of Information Photonics and Optical Metrology, Department of Physics, Indian Institute of Technology, Varanasi, India
16:15–16:30	Field of view enhancement of dynamic holographic displays using algorithms, devices, and systems: A review	Monika Rani CSIR-Central Scientific Instruments Organization, Sector 30C, Chandigarh, 160030, India
16:30–16:45	Holography with incoherent light	Akanksha Gautam Laboratory of Information Photonics and Optical Metrology, Department of Physics, Indian Institute of Technology, Varanasi, Uttar Pradesh, India
16:45–17:00	An asymmetric optical cryptosystem using physically unclonable functions in the Fresnel domain	Vinny Cris M Department of Physics, SRM University-AP, Andhra Pradesh–522502, India
17:00–17:15	Techniques to expand the exit pupil of Maxwellian display: A review	Rajveer Kaur CSIR–Central Scientific Instruments Organization, Sector 30C, Chandigarh, 160030, India
17:15–17:30	Imaging incoherent target using Hadamard basis patterns	Tanushree Karmakar Laboratory of Information Photonics and Optical Metrology, Department of Physics, Indian Institute of Technology, Varanasi, Uttar Pradesh, India
18:15–18:20	Closing Remarks	
Day 2: Tuesday, 21 February 2023 ADVANCED MANUFACTURING		
09:00–09:15	Virtual conference window open	
09:15–09:35	Opening remarks and welcome note	Prof. Vijayakumar Anand Associate Professor, University of Tartu
09:40–10:25	Keynote—Ultra-short laser pulses as material synthesis and lithography tool	Prof. Saulius Juodkazis Deputy Director of Optical Sciences Center and Director of Nanotechnology, Swinburne University of Technology, Melbourne, Australia
Session 1—AM1. Advanced Manufacturing Chairs: Dr. Soon Hock Ng and Daniel Smith		
10:30–10:50	3D auxetic metamaterials as scaffolds for tissue engineering	Prof. Maria Farsari Research Director at FORTH/IESL, Greece
10:55–11:15	Heavy-duty and high-performance 3D micro-optics made by laser additive manufacturing	Prof. Mangirdas Malinauskas Group Leader, Laser Research Center, Vilnius University, Lithuania
11:20–11:40	Femtosecond laser printing of form birefringent polymeric nanostructures	Prof. Vygantas Mizeikis Professor, Department of Engineering and Research Institute of Electronics of Shizuoka University, Japan
11:45–12:05	Knobs for tuning the physical properties of atom-thin graphene layers	Prof. Manu Jaiswal Professor, Department of Physics, IIT Madras, India
12:10–12:30	Efficient generation of higher harmonics from nonlinear metasurfaces and its applications	Dr. Aravind P. Anthur Research scientist, quantum technology for engineering (QTE) department of institute of materials research and engineering (IMRE), A *STAR, Singapore

Table 4. *Cont.*

	Session 2—AM2. Advanced Manufacturing Chairs: Prof. Kaupo Kukli and Andrei Bleahu	
12:35–12:55	The Optical Drill: twisted beams	Dr. Darius Gailevicius Researcher, Faculty of Physics, Laser Research Center, Vilnius University, Lithuania
13:00–13:20	Photon-induced structural and surface engineering for advanced manufacturing of nanoscale material	Dr. Jagadeesh Suriyaprakash Researcher, Guangdong Provincial Key Laboratory of Nanophotonic Functional Materials and Devices, SCNU, China
13:25–13:45	Mirror-less Laser action enabled by Energy Transfer from a conjugated oligomer donor to Chromeno-quinoline acceptor	Prof. Saradh Prasad Reseacher, Department of Physics and Astronomy, College of Science, King SAUD University, Saudi Arabia
13:50–14:10	M2 factor of conically refracted Gaussian beams	Dr. Erko Jalviste Researcher, Biophysics Lab of Institute of Physics, University of Tartu, Estonia
	Poster Presentations Session Chairs: Prof. Kaupo Kukli and Dr. Soon Hock Ng	
	14:30–17:00 (10 min for presentation + 2 min for Questions)	
14:30–14:45	3D scaffolds via Multi-Photon Polymerization as a co-culture system for application in peripheral nervous system regeneration.	Antonis Kordas Institute of Electronic Structure and Laser, Foundation for Research and Technology-Hellas (FORTH-IESL), and Department of Materials Science and Technology, University of Crete, Greece
14:45–15:00	Investigation of Memristor-based neural networks on Pattern recognition	Gayatri Routhu K L Deemed to be University, Andhra Pradesh, India
15:00–15:15	Additive micro-/nano manufacturing of non-sensitized SZ2080TM employing femtosecond-laser VIS-light oscillator	Antanas Butkus Laser Research Center, Faculty of Physics, Vilnius University, Lithuania
15:15–15:30	An efficient designing of IIR filter for ECG signal classification using MATLAB	Manjula Nandi K L Deemed to be University, Andhra Pradesh, India
15:30–15:45	Fabrication and analysis of 3D low THz metamaterials	Savvas Papamakarios Department of Physics, University of Crete and Institute of Electronic Structure and Laser, Foundation for Research and Technology-Hellas (FORTH-IESL), Greece
15:45–16:00	Advanced driver fatigue detection by integration of OpenCV, DNN module and deep learning	Muzammil Parvez M K L Deemed to be University, Andhra Pradesh, India
16:00–16:15	Design and simulation of a low-power and high-speed Fast Fourier Transform for medical image compression	Ernest Ravindran R S K L Deemed to be University, Andhra Pradesh, India
16:15–16:30	Analysis of 22 nm memristor-based inverter and universal gates using ANN model	Kota Bhagya Chandrika K L Deemed to be University, Andhra Pradesh, India
16:30–16:45	Optimization of placement and routing for ALU using reinforcement learning algorithm	Eppala Shashi Kumar Reddy K L Deemed to be University, Andhra Pradesh, India
17:00–17:05	Closing Remarks	
	Day 3: Wednesday, 22 February 2023 INDUSTRY APPLICATIONS	
09:00–09:15	Virtual conference window open	
09:15–09:30	Opening remarks	Dr. Amudhavel Jayavel
09:30–09:40	Welcome note	Ms. Tiia Lillemaa
09:40–10:25	Keynote—Mapping the path from idea to economically scalable application in 3D imaging and sensing	Prof. Heli Valtna
	Session 1—IA1. Industry Applications Chairs: Dr. Amudhavel Jayavel and Andrei-Ioan Bleahu	
10:30–10:50	New medical imaging: Physics, medical need and commercial viability	Dr. Zoltan Vilagosh RMIT Melbourne and Swinburne universities, Australia
10:55–11:15	Holography and its significance in trauma management and anorectal surgeries—An eye-opener!	Dr. Scott Arockia Singh Chairman and Managing Director in Dr. Scotts Clinic, Nagercoil, Tamil Nadu, India, Consultant Laparoscopic in Dr. Jeyasekharan Hospital, Nagercoil, Tamil Nadu, India

Table 4. *Cont.*

11:20–11:40	Imaging pitfalls and diagnostic inhibitions in various imaging modalities of head and neck region	Dr. Durgadevi B Senior Lecturer, Dept of Oral medicine and maxillofacial radiology, Indira Gandhi institute of dental sciences, Sri Balaji vidyapeeth (Deemed) to be University, Pondicherry, India
11:45–12:05	Holography and its clinical implications in Urology—An overview	Dr. Deepak David Consultant, Department of Urology, Dr. Jeyasekharan Hospital and Nursing Home, Nagercoil, Tamil Nadu, India
	Session 2—IA2. Industry Applications Chairs: Dr. Amudhavel Jayavel and Shivasubramanian Gopinath	
12:10–12:30	Light for life—Optical spectroscopy in clinical settings	Dr. Shree Krishnamoorthy Researcher at BioPhotonics team with Prof. Stefan Andersson-Engels at Tyndall National Institute in Cork, Ireland
12:35–12:55	Challenges in Burns Management—A Burns Surgeon's expectation from Modern-day Technology	Dr. Mohammed Imran Khan Consultant and heads the Department of Plastic and Reconstructive Surgery at Grace Kennet Foundation Hospital and Burns Centre, Madurai, India.
13:00–13:20	Cone beam tomography—Challenges and optimization	Prof. Dr. Milling Tania Department of Orthodontics and dentofacial orthopedics Rajas Dental College and Hospital, Kavalkinaru, Tirunelveli, Tamil Nadu, India, Consultant Orthodontist in Darshan Dental and Orthodontic Clinic, Kanyakumari, Tamil Nadu, India
13:25–13:45	Holography in bladder cancer and Renal stones	Dr. Priyadarshini D Jeyasekharan Hospital, Nagercoil, Tamil Nadu, India
	Poster Presentations Session Chairs: Dr. Amudhavel Jayavel and Dr. Viktor Palm	
	14:00–17:00 (10 min for presentation + 2 min for Questions)	
14:00–14:15	Study of a pH-sensitive hologram for biosensing applications	Komal Sharma CSIR-Central Scientific Instruments Organisation, Chandigarh, India and Academy of Scientific and Innovative Research (AcSIR), Ghaziabad, India
14:15–14:30	Enhancing phase measurement by a factor of two in the Stokes correlation	Amit Yadav Laboratory of Information Photonics and Optical Metrology, Department of Physics, Indian Institute of Technology, Varanasi, India
14:30–14:45	Hologram opens new learning door for postgraduate students—An academic view point	Dr. Thivagar M Jayanagar general hospital, Bengaluru, India
14:45–15:00	Design of low-power phase locked loop	Chandra Keerthi Pothina Department of Electronics and Communication, Koneru Lakshmaiah Education Foundation, Guntur India
15:00–15:15	Holographic technology as a new pain-reducing solution for Children	Anneli Kolk Faculty of Medicine, University of Tartu and Department of Pediatrics and Neurology, Tartu University Hospital Children's clinic
15:15–15:30	Verification of SoC using advanced verification methodology	Pranuti Pamula K L Deemed to be University, Andhra Pradesh, India
15:30–15:45	Correction of focusing errors of a refractive lens Using the Lucy–Richardson–Rosen algorithm	Andrei Bleahu CIPHR Group, Institute of Physics, University of Tartu
15:45–16:00	Implementation of content-based image retrieval using artificial neural networks	Sarath Yenigalla K L Deemed to be University, Andhra Pradesh, India)
16:00–16:15	Design and simulation of a low-power and high-speed Fast Fourier transform for medical image compression	Ernest Ravindran R S K L Deemed to be University, Andhra Pradesh, India
16:15–16:30	Best poster presentations announcement for days 1, 2 and 3	
16:30–16:45	Closing remarks	
	End of the conference	

5. Proceedings and Certificates

The accepted and presented submissions of 27 articles were published in the MDPI (Multidisciplinary Digital Publishing Institute, Basel, Switzerland) journal *Engineering Proceedings—Eng. Proc.*, 2023, HMAM2 https://www.mdpi.com/2673-4591/34/1 (accessed on 21 July 2023) [1–27]. Selected submissions were invited to the Special Issue on "Research in Computational Optics" in the MDPI journal *Photonics* (IF–2.536).

Participation e-certificates were given to all the participants online via the Sciforum platform through the following link—Download the participation certificate from the Sciforum https://hmam2.sciforum.net/-Log In > My certificates. The e-Certificates were awarded to all the invited speakers, keynote speakers, poster presentations, and winner certificates for poster presentations.

6. Materials from the Conference

The materials from the conference are available in the following sections in the Sciforum online platform event, HMAM2—Holography Meets Advanced Manufacturing: Welcome message from the chairs, event calls, event organizers, event chairs, event speakers, sessions, conference schedule, instructions for authors, list of submitted submission, sponsors and partners, poster gallery. https://hmam2.sciforum.net/.

7. Conclusions

Our three-day online conference on "Holography Meets Advanced Manufacturing (HMAM)" organized by the CIPHR group, Institute of Physics, University of Tartu, Estonia, and the Optical Sciences Center, Swinburne University of Technology, Australia, was successful. The conference gave the CIPHR team opportunities to network with researchers and medical doctors across the globe.

Author Contributions: Conceptualization, all the authors; methodology, all the authors; resources, all the authors; writing—original draft preparation, A.S.J.F.R.; writing—review and editing, all the authors; project administration, V.A., A.S.J.F.R., K.K., V.P., A.J. and S.J.; funding acquisition, V.A. and S.J. All authors have read and agreed to the published version of the manuscript.

Funding: This conference was funded by European Union's Horizon 2020 research and innovation programme grant agreement No. 857627 (CIPHR).

Acknowledgments: We would like to thank the Institute of Physics, University of Tartu, Estonia, and the Optical Sciences Center, Swinburne University of Technology, Australia, for supporting this event. The authors thank Tiia Lillemaa for her administrative support.

Conflicts of Interest: The authors declare no conflict of interest.

References

1. Boopathi, D. Imaging Pitfalls and Diagnostic Inhibitions in Various Advanced Head and Neck Imaging Modalities—Diagnostician's Perspective. *Eng. Proc.* **2023**, *34*, 1.
2. Rosen, J. Advanced Imaging Methods Using Coded Aperture Digital Holography. *Eng. Proc.* **2023**, *34*, 2.
3. Tahara, T.; Kozawa, Y.; Nakamura, T.; Matsuda, A.; Shimobaba, T. Incoherent Digital Holography for Multidimensional Motion Picture Imaging. *Eng. Proc.* **2023**, *34*, 3.
4. Yadav, A.; Sarkar, T.; Suzuki, T.; Singh, R.K. Enhancing Phase Measurement by a Factor of Two in the Stokes Correlation. *Eng. Proc.* **2023**, *34*, 4.
5. Gautam, A.; TS, A.; Naik, D.N.; Narayanmurthy, C.S.; Singh, R.; Singh, R.K. Holography with Incoherent Light. *Eng. Proc.* **2023**, *34*, 5.
6. Dodda, V.C.; Muniraj, I. Roles of Deep Learning in Optical Imaging. *Eng. Proc.* **2023**, *34*, 6.
7. Rathor, M.; Chaubey, S.K.; Singh, R.K. Digital Fourier Transform Holography Using a Beam Displacer. *Eng. Proc.* **2023**, *34*, 7.
8. Cris Mandapati, V.; Prabhakar, S.; Vardhan, H.; Kumar, R.; Reddy, S.G.; Sakshi; Singh, R.P. An Asymmetric Optical Cryptosystem Using Physically Unclonable Functions in the Fresnel Domain. *Eng. Proc.* **2023**, *34*, 8.
9. Routhu, G.; Phalguni Singh, N.; Raja, S.; Reddy, E.S.K. Investigation of Memristor-Based Neural Networks on Pattern Recognition. *Eng. Proc.* **2023**, *34*, 9.
10. Singh, R.K. Digital Polarization Holography: Challenges and Opportunities. *Eng. Proc.* **2023**, *34*, 10.

11. Rani, M.; Joshi, N.; Das, B.; Kumar, R. Field of View Enhancement of Dynamic Holographic Displays Using Algorithms, Devices, and Systems: A Review. *Eng. Proc.* **2023**, *34*, 11.
12. Pamula, P.; Gorthy, D.P.; Ngangbam, P.S.; Alagarsamy, A. Verification of SoC Using Advanced Verification Methodology. *Eng. Proc.* **2023**, *34*, 12.
13. Rajveer, K.; Raj, K. Techniques to Expand the Exit Pupil of Maxwellian Display: A Review. *Eng. Proc.* **2023**, *34*, 13.
14. Pothina, C.K.; Singh, N.P.; Prasanna, J.L.; Santhosh, C.; Kumar, M.R. Design of Efficient Phase Locked Loop for Low Power Applications. *Eng. Proc.* **2023**, *34*, 14.
15. Parvez, M.M.; Allanki, S.; Sudhagar, G.; RS, E.R.; Santosh, C.; Mohammed, A.B.; Muqeet, M.A. Advanced Driver Fatigue Detection by Integration of OpenCV DNN Module and Deep Learning. *Eng. Proc.* **2023**, *34*, 15.
16. Potcoava, M.; Mann, C.; Art, J.; Alford, S. Light Sheet Fluorescence Microscopy Using Incoherent Light Detection. *Eng. Proc.* **2023**, *34*, 16.
17. Karmakar, T.; Singh, R.; Singh, R.K. Imaging Incoherent Target Using Hadamard Basis Patterns. *Eng. Proc.* **2023**, *34*, 17.
18. Ramaswami Sachidanandan, E.R.; Phalguni Singh, N.; Gunda, S. Design and Simulation of a Low-Power and High-Speed Fast Fourier Transform for Medical Image Compression. *Eng. Proc.* **2023**, *34*, 18.
19. Shevkunov, I.; Katkovnik, V.; Egiazarian, K. Lensless Hyperspectral Phase Retrieval via Alternating Direction Method of Multipliers and Spectral Proximity Operators. *Eng. Proc.* **2023**, *34*, 19.
20. Gailevicius, D.; Zvirblis, R.; Malinauskas, M. Resilient Calcination Transformed Micro-Optics. *Eng. Proc.* **2023**, *34*, 20.
21. Thivagar, M. Hologram Opens a New Learning Door for Surgical Residents—An Academic View Point. *Eng. Proc.* **2023**, *34*, 21.
22. Sharma, K.; Heena; Mohanta, G.C.; Kumar, R. Study of a pH-Sensitive Hologram for Biosensing Applications. *Eng. Proc.* **2023**, *34*, 22.
23. Vilagosh, Z. New Medical Imaging, Physics, Medical Need and Commercial Viability. *Eng. Proc.* **2023**, *34*, 23.
24. Manjula, N.; Singh, N.P.; Babu, P.A. An Efficient Designing of IIR Filter for ECG Signal Classification Using MATLAB. *Eng. Proc.* **2023**, *34*, 24.
25. Yenigalla, S.C.; Rao, S.; Ngangbam, P.S. Implementation of Content-Based Image Retrieval Using Artificial Neural Networks. *Eng. Proc.* **2023**, *34*, 25.
26. Smith, D.; Ng, S.H.; Han, M.; Katkus, T.; Anand, V.; Juodkazis, S. Imaging with Diffractive Axicons Rapidly Milled on Sapphire by Femtosecond Laser Ablation. *Eng. Proc.* **2023**, *34*, 26.
27. Krishnamoorthy, S. Light for Life—Optical Spectroscopy in Clinical Settings. *Eng. Proc.* **2023**, *34*, 27.

Proceeding Paper

Advanced Imaging Methods Using Coded Aperture Digital Holography †

Joseph Rosen

School of Electrical and Computer Engineering, Ben-Gurion University of the Negev, P.O. Box 653, Beer-Sheva 8410501, Israel; rosenj@bgu.ac.il
† Presented at the International Conference on "Holography Meets Advanced Manufacturing", Online, 20–22 February 2023.

Abstract: Optical imaging has been utilized in nature and technology for decades. Recently, new methods of optical imaging assisted by computational imaging techniques have been proposed and demonstrated. We describe several new methods of three-dimensional optical imaging, from Fresnel incoherent correlation holography (FINCH) to interferenceless coded aperture correlation holography (COACH). FINCH and COACH are methods for recording digital holograms of a three-dimensional scene. However, COACH can be used for other incoherent and coherent optical applications. The possible applications for these imaging methods, ranging from a new generation of fluorescence microscopes to noninvasive imaging methods through a scattering medium, are mentioned.

Keywords: digital holography; incoherent holography; imaging systems; coded aperture

Citation: Rosen, J. Advanced Imaging Methods Using Coded Aperture Digital Holography. *Eng. Proc.* **2023**, *34*, 2. https://doi.org/10.3390/HMAM2-14122

Academic Editor: Kaupo Kukli

Published: 6 March 2023

1. Introduction

Coded aperture correlation holography (COACH), the main topic of this article, was proposed as a new technique of incoherent digital holography [1]. Hence, we begin this article with a brief history of imaging using holography [2], digital holography [3], and incoherent holography [4]. Since many holograms in the past and today have been recorded as the result of interference between two light waves, wave interference is the natural starting point. The phenomenon of optical two-wave interference has been well known since the first decade of the nineteenth century, when Thomas Young published his famous double slit experiment [5]. Young's experiment produces an interference pattern between two light waves, but this pattern is not considered a hologram because neither of the two interfering waves contains any image information.

The revolutionary transition from Young's interference pattern to a hologram occurred in 1948 in Dennis Gabor's pioneering work, presenting, for the first time, what is known today as the Gabor hologram [6]. This and similar holograms are recorded with two-wave interference between a wave carrying the object information and another wave called a reference wave, which does not contain any object information. However, the reference wave in the Gabor hologram passes through the observed object before the interference pattern between the beams is recorded on the photographic plate [6]. This type of hologram in which light from the object is used as a reference beam (although it does not contain any image information of the object but only the image background) is called a self-reference hologram [7]. Another distinct feature of the Gabor hologram, in contrast to the Young experiment, is the zero angle between the two interfering beams. A holographic recording system in which there is no angle between the reference and image beams is called an on-axis system. The Gabor hologram is also classified as a spatially coherent hologram because the light source illuminating the object is a point-like source. Holography, in general, is classified into coherent and incoherent holography depending on the light nature used for object illumination. Wave interference can be easily achieved with coherent light beams, but many imaging tasks are widely applicable only under incoherent illumination. In general, imaging systems under incoherent illumination have a frequency response called the modulation transfer function, with a larger spatial bandwidth than coherent systems with the same aperture dimensions [5]. Hence, the incoherent image usually has a higher image resolution than the coherent image. From now on, unless something else is explicitly said, "incoherent light" throughout this article refers to quasimonochromatic spatially incoherent light.

The next historical milestone in holography was the off-axis hologram proposed by Leith and Upatnieks in 1962 [8]. The recording configuration of this hologram is characterized by a nonzero angle between the image and reference beams, and consequently, the twin-image problem of the Gabor hologram is solved. The twin-image problem is the inability to extract the desired component representing the required image out of four components recorded on the raw hologram [9]. Because the twin-image problem is no longer a problem, the image of the observed object can be reconstructed from the off-axis hologram by illuminating it with a reference beam, and this image can be viewed clearly without interruptions by other light waves. The off-axis hologram is not a self-reference one, and from that aspect, it also differs from the Gabor hologram. In the aspect of spatial coherence of the illumination, the off-axis hologram is similar to Gabor's, as they are both considered spatially coherent holograms. The transition from the Gabor hologram to the off-axis hologram was easier with the invention of the laser, with its relatively high temporal coherence, since the optical path difference between the object and reference beams is not restricted as it is in the Gabor hologram.

Incoherent holograms have appeared since the mid-1960s [10,11], and all of them were based on different implementations of the self-interference principle [12]. The self-interference principle means that the light from each object point splits into two waves modulated differently before creating an interference pattern on the recording plane. According to this definition, a self-interference hologram is also a self-reference hologram because both interfering beams come from the same object. However, unlike self-interference, in a self-reference hologram, the reference beam does not contain image information. Under the self-interference principle, Bryngdahl and Lohmann suggested sorting interferometers for recording incoherent holograms into two types [13]. The first is radial shear, in which the observed image is replicated into two replications with two different scales. The other type is rotational shear, in which the observed image is also replicated into two versions, but in this case, one replication is rotated by some angle relative to the other replication. The entire holograms recorded using the self-interference principle are the stage in the evolutionary chain of holography in which both interfering waves carry the object's image. This new stage has practical meaning; under certain conditions, the self-interference principle leads to the violation of the Lagrange invariant [5], leading to better image resolution.

The next significant event in hologram history occurred in 1967, with the invention of the digital hologram by Goodman and Lawrence [14]. Digital holography is an indirect imaging technique where holograms are first acquired using a digital camera and then the image is reconstructed digitally through a computational algorithm [1,3]. Thus, digital holography is a two-step process that has some advantages over regular digital imaging. For example, a hologram can contain depth information of three-dimensional (3D) objects utilizing phase information encoded in the interference patterns between an object and the reference beams [1,3]. Other useful information recorded on a hologram might be the wavefront shape of the wave passing through the object, enabling quantitative phase imaging (QPI) [15]. The first digital hologram was coherent and recorded on a digital camera using an off-axis setup [14]. Another notable difference between this new digital hologram and those mentioned above is the transformation between the complex amplitudes on the object and the hologram planes. The two-dimensional (2D) Fourier transform was the transformation from the object to the hologram planes in the case of the Goodman–Lawrence hologram, thus indicating the type of hologram as a Fourier hologram. An optical (nondigital) Fourier hologram was proposed a few years before by Vander Lugt [16]. In 1997, Yamaguchi and Zhang recorded on-axis digital holograms in which the twin-image problem was solved by recording four different holograms of the coherently illuminated object and processing them in the computer in a procedure called phase shifting [17]. The transformation between the object and camera planes in the Yamaguchi–Zhang system follows Fresnel free-space propagation and, hence, this digital hologram is considered a Fresnel hologram [6,8,18].

In the field of incoherent digital holography, technology evolved, producing unexpected solutions. The minimal number of camera shots, one in the Goodman–Lawrence hologram [14] and four in the Yamaguchi–Zhang technique [17], was replaced by scanning techniques that do not make use of the self-interference principle. Under scanning techniques, there are two main methods of recording incoherent digital holograms of a general 3D scene. The more well-known method is optical scanning holography [19,20], in which the 3D object is scanned using an interference pattern between two spherical waves, and the reflected light is summed into a point detector. In optical scanning holography, the wave interference is between two spherical waves, neither of which carries any image. Moreover, the interference pattern is not recorded but is used as a detector of the object points' depth. The other scanning technique was implemented without wave interference and is a more computer-aided method, in which the hologram is generated from multiple view projections of the 3D scene [21,22]. Both methods are based on different processes of time-consuming scanning of the observed scene to yield a 2D correlation between an object and a 2D quadratic phase function.

The next landmark is that the required 2D correlation in Refs. [19–22] can be performed without scanning. Fresnel incoherent correlation holography (FINCH), published in 2007 [23], was a return to the principle of self-interference and was proposed as an alternative to the scanning-based holography methods mentioned above. Following the first FINCH, many other incoherent digital holograms were proposed, with most of them based on the self-interference principle [24–38]. Fourier incoherent single-channel holography (FISCH) [39] is a typical example of using the self-interference principle, but the obtained hologram, in this case, is a 2D cosine Fourier transform of the object. As mentioned above, in the entire holograms recorded using the self-interference principle, both interfering waves carry the object's image. However, the image information is never the same in both interferometer channels. In FINCH, the images are in focus at different distances from the aperture, while an infinite distance is also legitimate. In FISCH, one image is rotated by 180° around the origin of the image plane relative to the other image. In terms of the Bryngdahl–Lohmann analysis, FINCH is radial shear, and FISCH is rotational shear. An exceptional example of an incoherent digital hologram based on the self-reference rather than the self-interference principle was proposed by Pedrini et al. [40], but the energetic inefficiency of this hologram recorder probably prevented further developments in this direction.

2. Coded Aperture Correlation Holography (COACH)

COACH [41–43] is a new evolutionary stage in which one of the two replicated objects' images passes through a coded scattering mask, resulting in the camera plane being a convolution of the image with some chaotic function. The other image is in focus at an infinite distance from the aperture. According to the classifications mentioned above, COACH is radial shear and belongs to incoherent self-interference on-axis digital holography. A significant difference between COACH and Fresnel holograms is in the image reconstruction process. In the Fresnel case, the image at a distance of z is reconstructed through a correlation between the hologram and a quadratic phase function parameterized with z. On the other hand, in COACH, the 3D image is reconstructed through a correlation between the hologram and a library of point responses acquired in the system calibration. From the COACH stage, the technology surprisingly evolved to a system without two-wave interference following the discovery that 3D holographic imaging could be achieved with a single-beam configuration. The interferenceless COACH (I-COACH) [44] was found to be simpler and more efficient than the COACH with two-wave interference. I-COACH is considered a digital hologram because the digital matrix obtained from the observed scene contains the scene's 3D information, and the 3D image is reconstructed from the digital matrix in a similar way to interference-based digital holograms. Although there is no two-wave interference in I-COACH, it is classified as on-axis digital holography because the recording setup contains components that are all arranged along a single longitudinal axis. Using the interferenceless version of COACH has enabled adapting concepts from coded aperture imaging via X-ray [45], in which the observed image is replicated over a finite number of randomly distributed points. In other words, the point response of the system was modified from the continuous chaotic light distribution [41–44] to a chaotic ensemble of light dots [46]. Moreover, by integrating concepts from optical pattern recognition [47,48], the process of correlation-based image reconstruction was modified to what is known as nonlinear image reconstruction [49]. Because of the two modifications, the modified impulse response and the change in the reconstruction process, I-COACH's signal-to-noise ratio (SNR) has been improved significantly [50]. Other imaging properties, in addition to SNR, have also been treated in the framework of COACH research. The image resolutions of I-COACH have been improved via several different techniques [50–52]. Field-of-view (FOV) extension in I-COACH systems was addressed in Ref. [53] through a special calibration procedure. Ideas adapted from axial beam shaping have enabled engineering the depth of field (DOF) of an I-COACH system [54]. Sectioning the imaging space or, in other words, removing the out-of-focus background from the resulting picture, was demonstrated through point spread functions of tilted pseudo-nondiffracting beams in I-COACH [55]. Color imaging using various I-COACH systems has been treated in I-COACH [56] and in a setup with a quasirandom lens [57].

COACH can implement several applications in addition to the initial and widely used application of 3D holographic imaging. For example, noninvasive imaging through scattering layers can be more efficient if the light emitted from the scattering layer is modulated by a phase aperture, as demonstrated in Ref. [58]. Another application is imaging using telescopes with an annular aperture, which is a way to reduce the weight of space-based telescopes [59]. The images produced by such telescopes might be clearer and sharper using COACH [60]. Imaging with a synthetic aperture system is another example that enables better image resolution without changing the physical size of the optical aperture [61]. COACH can image targets with an incoherent synthetic aperture, with the advantage that the relatively small apertures move only along the perimeter of the relatively large synthetic aperture [62]. Although interferenceless imaging systems are simpler and more power efficient than systems with wave interference, the latter systems still have an important role in the technology, and the annular synthetic aperture [62,63] is an example of using two-wave interference between beams reflected from a pair of sub-apertures located along the aperture perimeter. More details about these advances and others of COACH and I-COACH can be found in two review articles [64–66]. The scheme in Figure 1 summarizes the holography history, as described above, where the blue arrows indicate the flow and influence of various ideas. The next natural step was to explore the new COACH concept in the area of coherent holography. In addition to 3D imaging, QPI is another main application for coherent holography. Thus, 3D imaging under coherent light using I-COACH was demonstrated in Ref. [67] but without phase imaging capability. QPI could not be performed using I-COACH, but various ways to implement QPI using phase apertures with [68] and without [69] two-wave interference and with [70] and without [71] using self-reference holography were found. Specifically, COACH's concepts have been integrated into a Mach–Zehnder interferometer [71], with the benefit of a broader FOV than a conventional QPI interferometer. A closely related technique of QPI is wavefront sensing, where a COACH-based Shack–Hartmann wavefront sensor was proposed recently [72], with the advantage of higher accuracy over the conventional Shack–Hartmann wavefront sensor.

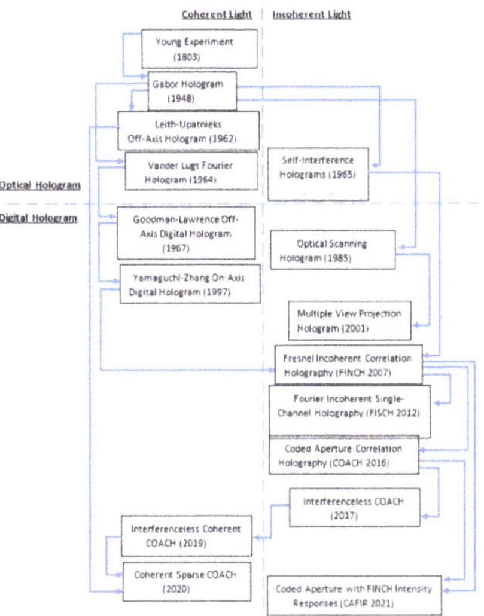

Figure 1. Scheme of holography history as described in the text. The blue arrows indicate the flow and influence of the various ideas.

3. Summary

The main benefit of sparse COACH is the ability to control the SNR and the visibility of the reconstructed image through the sparsity and complexity of the PSHs. However, in Ref. [73], we employ the sparse response of COACH to merge the imaging merits of FINCH and COACH into a single holographic system. In the apparatus of Ref. [73], the combination is achieved by granting the sparse COACH a response of a FINCH-type self-interference mechanism so that neither of the resolution types is compromised. In other words, the recently proposed imaging method integrates advantages from both the FINCH and COACH techniques, such that this hybrid system has the improved lateral resolution of FINCH with the same axial resolution of COACH.

The development of holography has not ended, and from time to time, a new improvement is published, so this article is only an interim summary of the field. However, the rapid development of COACH and other methods of phase aperture digital holography in incoherent and coherent optics might make this review a useful source for the holography community.

Funding: This study is supported by The Israel Innovation Authority under MAGNET program No. 79555.

Institutional Review Board Statement: Not applicable.

Informed Consent Statement: Not applicable.

Data Availability Statement: Not applicable.

Conflicts of Interest: The author declares no conflict of interest.

References

1. Tahara, T.; Zhang, Y.; Rosen, J.; Anand, V.; Cao, L.; Wu, J.; Koujin, T.; Matsuda, A.; Ishii, A.; Kozawa, Y.; et al. Roadmap of incoherent digital holography. *Appl. Phys. B* **2022**, *128*, 193. [CrossRef]
2. Sheridan, J.T.; Kostuk, R.K.; Gil, A.F.; Wang, Y.; Lu, W.; Zhong, H.; Tomita, Y.; Neipp, C.; Francés, J.; Gallego, S.; et al. Roadmap on holography. *J. Opt.* **2020**, *22*, 123002. [CrossRef]
3. Javidi, B.; Carnicer, A.; Anand, A.; Barbastathis, G.; Chen, W.; Ferraro, P.; Goodman, J.; Horisaki, R.; Khare, K.; Kujawinska, M.; et al. Roadmap on digital holography. *Opt. Express* **2021**, *29*, 35078–35118. [CrossRef] [PubMed]

4.   Braat, J.; Lowenthal, S. Short-exposure spatially incoherent holography with a plane-wave illumination. *J. Opt. Soc. Am.* **1973**, *63*, 388–390. [CrossRef]
5.   Born, M.; Wolf, E. *Principles of Optics*, 7th ed.; Cambridge University Press: Cambridge, MA, USA, 2019.
6.   Gabor, D. A new microscopic principle. *Nature* **1948**, *161*, 777–778. [CrossRef] [PubMed]
7.   Ma, L.; Wang, Y. Self-reference hologram. *Proc. SPIE* **2008**, *6837*, 68371.
8.   Leith, E.; Upatnieks, J. Reconstructed Wavefronts and Communication Theory. *J. Opt. Soc. Am.* **1962**, *52*, 1123–1130. [CrossRef]
9.   Stoykova, E.; Kang, H.; Park, J. Twin-image problem in digital holography-a survey (Invited Paper). *Chin. Opt. Lett.* **2014**, *12*, 060013. [CrossRef]
10.   Lohmann, A.W. Wavefront reconstruction for incoherent objects. *J. Opt. Soc. Am.* **1965**, *55*, 1555–1556. [CrossRef]
11.   Stroke, G.W.; Restrick, R.C. Holography with spatially noncoherent light. *Appl. Phys. Lett.* **1965**, *7*, 229–231. [CrossRef]
12.   Rosen, J.; Vijayakumar, A.; Kumar, M.; Rai, M.R.; Kelner, R.; Kashter, Y.; Bulbul, A.; Mukherjee, S. Recent advances in self-interference incoherent digital holography. *Adv. Opt. Photon.* **2019**, *11*, 1–66. [CrossRef]
13.   Bryngdahl, O.; Lohmann, A. Variable magnification in incoherent holography. *Appl. Opt.* **1970**, *9*, 231–232. [CrossRef] [PubMed]
14.   Goodman, J.W.; Lawrence, R.W. Digital image formation from electronically detected holograms. *Appl. Phys. Lett.* **1967**, *11*, 77–79. [CrossRef]
15.   Balasubramani, V.; Kujawińska, M.; Allier, C.; Anand, V.; Cheng, C.-J.; Depeursinge, C.; Hai, N.; Juodkazis, S.; Kalkman, J.; Kuś, A.; et al. Roadmap on Digital Holography-Based Quantitative Phase Imaging. *J. Imaging* **2021**, *7*, 252. [CrossRef] [PubMed]
16.   Vander Lugt, A. Signal detection by complex spatial filtering. *IEEE Trans. Inf. Theory* **1964**, *IT-10*, 139–145. [CrossRef]
17.   Yamaguchi, I.; Zhang, T. Phase-shifting digital holography. *Opt. Lett.* **1997**, *22*, 1268–1270. [CrossRef]
18.   Guel-Sandoval, S.; Ojeda-Castañeda, J. Quasi-Fourier transform of an object from a Fresnel hologram. *Appl. Opt.* **1979**, *18*, 950–951. [CrossRef] [PubMed]
19.   Poon, T.-C. Scanning holography and two-dimensional image processing by acousto-optic two-pupil synthesis. *J. Opt. Soc. Am. A* **1985**, *2*, 521–527. [CrossRef]
20.   Poon, T.-C.; Doh, K.B.; Schilling, B.W.; Wu, M.H.; Shinoda, K.K.; Suzuki, Y. Three-dimensional microscopy by optical scanning holography. *Opt. Eng.* **1995**, *34*, 1338–1344. [CrossRef]
21.   Li, Y.; Abookasis, D.; Rosen, J. Computer-generated holograms of three-dimensional realistic objects recorded without wave interference. *Appl. Opt.* **2001**, *40*, 2864–2870. [CrossRef]
22.   Shaked, N.T.; Katz, B.; Rosen, J. Review of three-dimensional imaging by multiple-viewpoint-projection based methods. *Appl. Opt.* **2009**, *48*, H120–H136. [CrossRef] [PubMed]
23.   Rosen, J.; Brooker, G. Digital spatially incoherent Fresnel holography. *Opt. Lett.* **2007**, *32*, 912–914. [CrossRef] [PubMed]
24.   Rosen, J.; Siegel, N.; Brooker, G. Theoretical and experimental demonstration of resolution beyond the Rayleigh limit by FINCH fluorescence microscopic imaging. *Opt. Express* **2011**, *19*, 26249–26268. [CrossRef]
25.   Brooker, G.; Siegel, N.; Wang, V.; Rosen, J. Optimal resolution in Fresnel incoherent correlation holographic fluorescence microscopy. *Opt. Express* **2011**, *19*, 5047–5062. [CrossRef]
26.   Rosen, J.; Brooker, G. Fresnel incoherent correlation holography (FINCH): A review of research. *Adv. Opt. Technol.* **2012**, *1*, 151–169. [CrossRef]
27.   Brooker, G.; Siegel, N.; Rosen, J.; Hashimoto, N.; Kurihara, M.; Tanabe, A. In-line FINCH super resolution digital holographic fluorescence microscopy using a high efficiency transmission liquid crystal GRIN lens. *Opt. Lett.* **2013**, *38*, 5264–5267. [CrossRef]
28.   Kashter, Y.; Vijayakumar, A.; Miyamoto, Y.; Rosen, J. Enhanced super resolution using Fresnel incoherent correlation holography with structured illumination. *Opt. Lett.* **2016**, *41*, 1558–1561. [CrossRef]
29.   Tahara, T.; Kanno, T.; Arai, Y.; Ozawa, T. Single-shot phase-shifting incoherent digital holography. *J. Opt.* **2017**, *19*, 065705. [CrossRef]
30.   Nobukawa, T.; Muroi, T.; Katano, Y.; Kinoshita, N.; Ishii, N. Single-shot phase-shifting incoherent digital holography with multiplexed checkerboard phase gratings. *Opt. Lett.* **2018**, *43*, 1698–1701. [CrossRef]
31.   Vijayakumar, A.; Katkus, T.; Lundgaard, S.; Linklater, D.; Ivanova, E.P.; Ng, S.H.; Juodkazis, S. Fresnel incoherent correlation holography with single camera shot. *Opto-Electron. Adv.* **2020**, *3*, 200004. [CrossRef]
32.   Marar, A.; Kner, P. Three-dimensional nanoscale localization of point-like objects using self-interference digital holography. *Opt. Lett.* **2020**, *45*, 591–594. [CrossRef]
33.   Potcoava, M.; Mann, C.; Art, J.; Alford, S. Spatio-temporal performance in an incoherent holography lattice light-sheet microscope (IHLLS). *Opt. Express* **2021**, *29*, 23888–23901. [CrossRef] [PubMed]
34.   Tahara, T.; Koujin, T.; Matsuda, A.; Ishii, A.; Ito, T.; Ichihashi, Y.; Oi, R. Incoherent color digital holography with computational coherent superposition for fluorescence imaging [Invited]. *Appl. Opt.* **2021**, *60*, A260–A267. [CrossRef] [PubMed]
35.   Anand, V.; Katkus, T.; Ng, S.H.; Juodkazis, S. Review of Fresnel incoherent correlation holography with linear and nonlinear correlations [Invited]. *Chin. Opt. Lett.* **2021**, *19*, 020501. [CrossRef]
36.   Tahara, T.; Ito, T.; Ichihashi, Y.; Oi, R. Multiwavelength three-dimensional microscopy with spatially incoherent light, based on computational coherent superposition. *Opt. Lett.* **2020**, *45*, 2482–2485. [CrossRef]
37.   Hara, T.; Tahara, T.; Ichihashi, Y.; Oi, R.; Ito, T. Multiwavelength-multiplexed phase-shifting incoherent color digital holography. *Opt. Express* **2020**, *28*, 10078–10089. [CrossRef]

38. Rosen, J.; Alford, S.; Anand, V.; Art, J.; Bouchal, P.; Bouchal, Z.; Erdenebat, M.-U.; Huang, L.; Ishii, A.; Juodkazis, S.; et al. Roadmap on recent progress in FINCH technology. *J. Imaging* **2021**, *7*, 197. [CrossRef]
39. Kelner, R.; Rosen, J. Spatially incoherent single channel digital Fourier holography. *Opt. Lett.* **2012**, *37*, 3723–3725. [CrossRef]
40. Pedrini, G.; Li, H.; Faridian, A.; Osten, W. Digital holography of self-luminous objects by using a Mach–Zehnder setup. *Opt. Lett.* **2012**, *37*, 713–715. [CrossRef]
41. Vijayakumar, A.; Kashter, Y.; Kelner, R.; Rosen, J. Coded aperture correlation holography—A new type of incoherent digital holograms. *Opt. Express* **2016**, *24*, 12430–12441. [CrossRef]
42. Vijayakumar, A.; Rosen, J. Spectrum and space resolved 4D imaging by coded aperture correlation holography (COACH) with diffractive objective lens. *Opt. Lett.* **2017**, *42*, 947–950. [CrossRef] [PubMed]
43. Vijayakumar, A.; Kashter, Y.; Kelner, R.; Rosen, J. Coded aperture correlation holography system with improved performance [Invited]. *Appl. Opt.* **2017**, *56*, F67–F77. [CrossRef] [PubMed]
44. Vijayakumar, A.; Rosen, J. Interferenceless coded aperture correlation holography—A new technique for recording incoherent digital holograms without two-wave interference. *Opt. Express* **2017**, *25*, 13883–13896. [CrossRef] [PubMed]
45. Ables, J.G. Fourier transform photography: A new method for X-ray astronomy. *Proc. Astron. Soc. Aust.* **1968**, *1*, 172–173. [CrossRef]
46. Rai, M.R.; Rosen, J. Noise suppression by controlling the sparsity of the point spread function in interferenceless coded aperture correlation holography (I-COACH). *Opt. Express* **2019**, *27*, 24311–24323. [CrossRef] [PubMed]
47. Fleisher, M.; Mahlab, U.; Shamir, J. Entropy optimized filter for pattern recognition. *Appl. Opt.* **1990**, *29*, 2091–2098. [CrossRef]
48. Kotzer, T.; Rosen, J.; Shamir, J. Multiple-object input in nonlinear correlation. *Appl. Opt.* **1993**, *32*, 1919–1932. [CrossRef]
49. Rai, M.R.; Vijayakumar, A.; Rosen, J. Nonlinear adaptive three-dimensional imaging with interferenceless coded aperture correlation holography (I-COACH). *Opt. Express* **2018**, *26*, 18143–18154. [CrossRef]
50. Rai, M.R.; Rosen, J. Resolution-enhanced imaging using interferenceless coded aperture correlation holography with sparse point response. *Sci. Rep.* **2020**, *10*, 5033. [CrossRef]
51. Rai, M.R.; Vijayakumar, A.; Ogura, Y.; Rosen, J. Resolution enhancement in nonlinear interferenceless COACH with point response of subdiffraction limit patterns. *Opt. Express* **2019**, *27*, 391–403. [CrossRef]
52. Rai, M.R.; Vijayakumar, A.; Rosen, J. Superresolution beyond the diffraction limit using phase spatial light modulator between incoherently illuminated objects and the entrance of an imaging system. *Opt. Lett.* **2019**, *44*, 1572–1575. [CrossRef] [PubMed]
53. Rai, M.R.; Vijayakumar, A.; Rosen, J. Extending the field of view by a scattering window in an I-COACH system. *Opt. Lett.* **2018**, *43*, 1043–1046. [CrossRef] [PubMed]
54. Rai, M.R.; Rosen, J. Depth-of-field engineering in coded aperture imaging. *Opt. Express* **2021**, *29*, 1634–1648. [CrossRef] [PubMed]
55. Hai, N.; Rosen, J. Single viewpoint tomography using point spread functions of tilted pseudo-nondiffracting beams in interferenceless coded aperture correlation holography with nonlinear reconstruction. *SSRN* **2023**. [CrossRef]
56. Dubey, N.; Kumar, R.; Rosen, J. Multi-wavelength imaging with extended depth of field using coded apertures and radial quartic phase functions. 2023; *submitted for publication*.
57. Anand, V.; Ng, S.; Katkus, T.; Juodkazis, S. White light three-dimensional imaging using a quasi-random lens. *Opt. Express* **2021**, *29*, 15551–15563. [CrossRef]
58. Mukherjee, S.; Vijayakumar, A.; Rosen, J. Spatial light modulator aided noninvasive imaging through scattering layers. *Sci. Rep.* **2019**, *9*, 17670. [CrossRef]
59. Rey, J.J.; Wirth, A.; Jankevics, A.; Landers, F.; Rohweller, D.; Chen, C.B.; Bronowicki, A. A deployable, annular, 30 m telescope, space-based observatory. *Proc. SPIE* **2014**, *9143*, 18.
60. Bulbul, A.; Rosen, J. Partial aperture imaging system based on sparse point spread holograms and nonlinear cross-correlations. *Sci. Rep.* **2020**, *10*, 21983. [CrossRef]
61. Merkle, F. Synthetic-aperture imaging with the European Very Large Telescope. *J. Opt. Soc. Am. A* **1988**, *5*, 904–913. [CrossRef]
62. Bulbul, A.; Vijayakumar, A.; Rosen, J. Superresolution far-field imaging by coded phase reflectors distributed only along the boundary of synthetic apertures. *Optica* **2018**, *5*, 1607–1616. [CrossRef]
63. Desai, J.P.; Kumar, R.; Rosen, J. Optical incoherent imaging using annular synthetic aperture with superposition of phase-shifted optical transfer functions. *Opt. Lett.* **2022**, *47*, 4012–4015. [CrossRef]
64. Rosen, J.; Anand, V.; Rai, M.R.; Mukherjee, S.; Bulbul, A. Review of 3D imaging by coded aperture correlation holography (COACH). *Appl. Sci.* **2019**, *9*, 605. [CrossRef]
65. Rosen, J.; Hai, N.; Rai, M.R. Recent progress in digital holography with dynamic diffractive phase apertures [Invited]. *Appl. Opt.* **2022**, *61*, B171–B180. [CrossRef]
66. Rosen, J.; Bulbul, A.; Hai, N.; Rai, M.R. Coded Aperture Correlation Holography (COACH)—A Research Journey from 3D Incoherent Optical Imaging to Quantitative Phase Imaging. In *Holography–Recent Advances and Applications*, 1st ed.; Rosen, J., Ed.; IntechOpen: London, UK, 2023.
67. Hai, N.; Rosen, J. Interferenceless and motionless method for recording digital holograms of coherently illuminated 3D objects by coded aperture correlation holography system. *Opt. Express* **2019**, *27*, 24324–24339. [CrossRef]
68. Hai, N.; Rosen, J. Phase contrast-based phase retrieval: A bridge between qualitative phase contrast and quantitative phase imaging by phase retrieval algorithms. *Opt. Lett.* **2020**, *45*, 5812–5815. [CrossRef]

69. Hai, N.; Kumar, R.; Rosen, J. Single-shot TIE using polarization multiplexing (STIEP) for quantitative phase imaging. *Opt. Lasers Eng.* **2022**, *151*, 106912. [CrossRef]
70. Hai, N.; Rosen, J. Single-plane and multiplane quantitative phase imaging by self-reference on-axis holography with a phase-shifting method. *Opt. Express* **2021**, *29*, 24210–24225. [CrossRef]
71. Hai, N.; Rosen, J. Coded aperture correlation holographic microscope for single-shot quantitative phase and amplitude imaging with extended field of view. *Opt. Express* **2020**, *28*, 27372–27386. [CrossRef]
72. Dubey, N.; Kumar, R.; Rosen, J. COACH-based Shack-Hartmann wavefront sensor with an array of phase coded masks. *Opt. Express* **2021**, *29*, 31859–31874. [CrossRef]
73. Bulbul, A.; Hai, N.; Rosen, J. Coded aperture correlation holography (COACH) with a superior lateral resolution of FINCH and axial resolution of conventional direct imaging systems. *Opt. Express* **2021**, *29*, 42106–42118. [CrossRef]

Proceeding Paper

Incoherent Digital Holography for Multidimensional Motion Picture Imaging [†]

Tatsuki Tahara [1],*, Yuichi Kozawa [2], Tomoya Nakamura [3], Atsushi Matsuda [4] and Tomoyoshi Shimobaba [5]

[1] Applied Electromagnetic Research Center, Radio Research Institute, National Institute of Information and Communications Technology, 4-2-1 Nukuikitamachi, Tokyo 184–8795, Japan

[2] Institute of Multidisciplinary Research for Advanced Materials, Tohoku University, 2-1-1 Katahira, Aoba-ku, Sendai 980-8577, Japan; y.kozawa@tohoku.ac.jp

[3] SANKEN, Osaka University, 8-1 Mihogaoka, Osaka 567-0047, Japan; nakamura@am.sanken.osaka-u.ac.jp

[4] Advanced ICT Research Institute Kobe, National Institute of Information and Communications Technology, 588-2 Iwaoka, Iwaoka-cho, Nishi-ku, Kobe 651-2492, Japan; a.matsuda@nict.go.jp

[5] Graduate School of Engineering, Chiba University, 1-33 Yayoi-cho, Inage-ku, Chiba 263-8522, Japan; shimobaba@faculty.chiba-u.jp

* Correspondence: tahara@nict.go.jp

† Presented at the International Conference on "Holography Meets Advanced Manufacturing", Online, 20–22 February 2023.

Abstract: Incoherent digital holography (IDH) is a technique used to obtain a three-dimensional (3D) image of spatially incoherent light diffracted from an object as an incoherent hologram. Color holographic 3D motion picture imaging of daily-use light at the frame rate of a color polarization imaging camera can be achieved by the combination of IDH and single-shot phase-shifting interferometry. We show experimental results for color 3D motion picture imaging in this proceedings article.

Keywords: incoherent digital holography; 3D motion picture imaging; color 3D imaging

1. Introduction

Incoherent digital holography (IDH) [1–14] is a three-dimensional (3D) image-sensing technique using interference of light and spatially incoherent light. Interference fringe images that contain 3D information about an object are obtained, even for spatially and temporally incoherent light, by generating two waves from an object wave and utilizing self-interference. A digital hologram of the daily use of light can be obtained using IDH, and the applications of this technology to fluorescence microscopy [15–20] and 3D imaging [21–28] have been actively researched. Full-color holographic 3D imaging using IDH has been shown to be possible, even for sunlight [5,28].

IDH has the ability for simultaneous imaging of multidimensional information such as a 3D image, multiple wavelengths [18–20,29–31], a state of polarization, and a variety of types of light [32]. Holographic quantitative phase imagers can be constructed using a small light emitting diode (LED) [33,34]. High-speed image sensing and robustness against external vibrations are important factors when constructing a multidimensional IDH system. Single-shot IDH [8–14] performed using single-shot phase-shifting [35–37] has been proposed as an IDH technique capable of satisfying the factors at play. In most of this IDH technique, holographic 3D imaging can be carried out using single-shot exposure of a polarization image sensor and a single-path interferometer. In this publication, we briefly introduce these holography techniques and the multidimensional imaging possible with this IDH technique.

Citation: Tahara, T.; Kozawa, Y.; Nakamura, T.; Matsuda, A.; Shimobaba, T. Incoherent Digital Holography for Multidimensional Motion Picture Imaging. *Eng. Proc.* **2023**, *34*, 3. https://doi.org/10.3390/HMAM2-14153

Academic Editor: Vijayakumar Anand

Published: 13 March 2023

2. Digital Holography Systems Adopting Single-Shot Phase-Shifting Interferometry for Multidimensional Motion Picture Imaging

Figure 1 illustrates two types of single-shot single-path digital holography (DH) systems: a single-shot full-color IDH system with birefringent materials [27,28] and a single-shot DH system for quantitative phase imaging with LED light [34]. A single-path self-interference or self-reference interferometer is adopted to the IDH systems. In these IDH systems, two waves are generated from a wave diffracted from an object. Coherence length should be considered when obtaining a digital hologram of natural light because of its low temporal coherence. These single-path IDH systems are designed to generate an interference fringe image of temporally incoherent light with high visibility. Differences in optical path length between the two waves can be carefully adjusted using polarimetric optical elements. Figure 1a depicts how a full-color hologram of natural light is obtained with a single exposure of a color polarization image sensor and with single-shot phase shifting. The DH system shown in Figure 1b is based on self-reference interferometer and has improved the depth resolution of DH with an LED light in comparison to self-interference IDH [33,34].

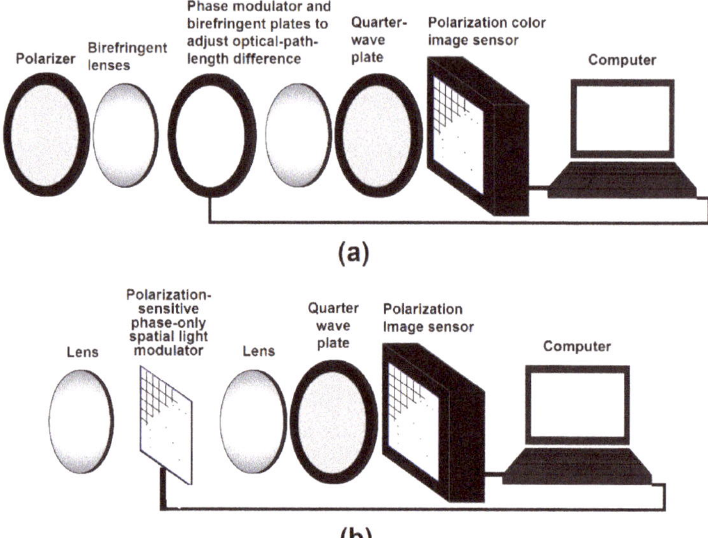

Figure 1. Single-shot single-path DH systems. (**a**) Single-shot full-color IDH system with birefringent materials [27,28]; (**b**) Self-reference DH system with a commonly used light source [34].

Figure 2 shows photographs of the constructed single-shot full-color IDH system and an example of experimental results obtained with the constructed IDH system. This IDH system can be used on a wood table to record a full-color hologram of objects illuminated by sunlight [27]. One study obtained a full-color holographic image from a single recorded hologram using the constructed IDH system and an RGB-LED, as shown in Figure 2b,c. Video-rate full-color holographic 3D motion picture imaging has also been experimentally demonstrated via the setup [28].

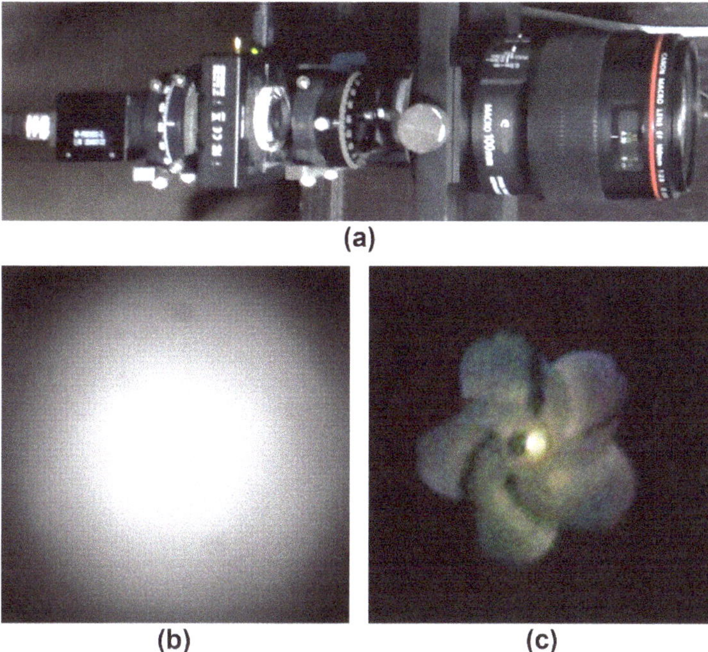

Figure 2. An optical implementation and an example of experimental results. (**a**) Constructed single-shot full-color IDH system with birefringent materials with a camera lens, termed "Holocamera"; (**b**) A recorded hologram with a holocamera; (**c**) the image reconstructed from (**a**).

3. Conclusions

We have briefly introduced a single-shot single-path IDH with which to perform multidimensional imaging. As another study remarked, the limitation of the measurement accuracy in interferometry and holography can be quantitatively evaluated based on the theory of quantum optics [38]. Algorithms and architectures for high-speed image reconstruction are also highly required for use in real-time measurement [39–41]. We will show experimental results for multidimensional holographic imaging of incoherent light, such as sunlight and LED light, in the presentation.

Author Contributions: All authors contributed equally. This manuscript was compiled by T.T. All authors have read and agreed to the published version of the manuscript.

Funding: This study was partially supported by Precursory Research for Embryonic Science and Technology (PRESTO) (JPMJPR16P8), Japan Society for the Promotion of Science (JSPS) (JP19H03202), The Cooperative Research Program of "Network Joint Research Center for Materials and Devices" (No. 20224020 and 20234030), and The Mitsubishi Foundation (202111007).

Institutional Review Board Statement: Not applicable.

Informed Consent Statement: Not applicable.

Data Availability Statement: Data is contained within the article.

Acknowledgments: We thank co-workers and co-authors of our articles related to IDH.

Conflicts of Interest: The authors declare no conflict of interests.

References

1. Poon, T.-C.; Korpel, A. Optical transfer function of an acousto-optic heterodyning image processor. *Opt. Lett.* **1979**, *4*, 317–319. [CrossRef] [PubMed]
2. Itoh, K.; Inoue, T.; Yoshida, T.; Ichioka, Y. Interferometric super-multispectral imaging. *Appl. Opt.* **1990**, *29*, 1625–1630. [CrossRef] [PubMed]
3. Mugnier, L.M.; Sirat, G.Y. On-axis conoscopic holography without a conjugate image. *Opt. Lett.* **1992**, *17*, 294–296. [CrossRef] [PubMed]
4. Rosen, J.; Brooker, G. Digital spatially incoherent Fresnel holography. *Opt. Lett.* **2007**, *32*, 912–914. [CrossRef]
5. Kim, M.K. Full color natural light holographic camera. *Opt. Express* **2013**, *21*, 9636–9642. [CrossRef]
6. Vijayakumar, A.; Kashter, Y.; Kelner, R.; Rosen, J. Coded aperture correlation holography—A new type of incoherent digital holograms. *Opt. Express* **2016**, *24*, 12430–12441. [CrossRef]
7. Wu, J.; Zhang, H.; Zhang, W.; Jin, G.; Cao, L.; Barbastathis, G. Singleshot lensless imaging with Fresnel zone aperture and incoherent illumination. *Light Sci. Appl.* **2020**, *9*, 53. [CrossRef]
8. Rosen, J.; Vijayakumar, A.; Kumar, M.; Rai, M.R.; Kelner, R.; Kashter, Y.; Bulbul, A.; Mukherjee, S. Recent advances in self-interference incoherent digital holography. *Adv. Opt. Photon.* **2019**, *11*, 1–66. [CrossRef]
9. Hong, J.; Kim, M.K. Overview of techniques applicable to self-interference incoherent digital holography. *J. Eur. Opt. Soc. Rapid Publ.* **2013**, *8*, 13077. [CrossRef]
10. Liu, J.-P.; Tahara, T.; Hayasaki, Y.; Poon, T.-C. Incoherent digital holography: A review. *Appl. Sci.* **2018**, *8*, 143. [CrossRef]
11. Tahara, T.; Quan, X.; Otani, R.; Takaki, Y.; Matoba, O. Digital holography and its multidimensional imaging applications: A review. *Microscopy* **2018**, *67*, 55–67. [CrossRef] [PubMed]
12. Rosen, J.; Alford, S.; Anand, V.; Art, J.; Bouchal, P.; Bouchal, Z.; Erdenebat, M.U.; Huang, L.; Ishii, A.; Juodkazis, S.; et al. Roadmap on recent progress in FINCH technology. *J. Imaging* **2021**, *7*, 197. [CrossRef] [PubMed]
13. Tahara, T. Review of incoherent digital holography: Applications to multidimensional incoherent digital holographic microscopy and palm-sized digital holographic recorder-holosensor. *Front. Photonics* **2022**, *2*, 829139. [CrossRef]
14. Tahara, T.; Zhang, Y.; Rosen, J.; Anand, V.; Cao, L.; Wu, J.; Koujin, T.; Matsuda, A.; Ishii, A.; Kozawa, Y.; et al. Roadmap of incoherent digital holography. *Appl. Phys. B* **2022**, *128*, 193. [CrossRef]
15. Schilling, B.W.; Poon, T.-C.; Indebetouw, G.; Storrie, B.; Shinoda, K.; Suzuki, Y.; Wu, M.H. Three-dimensional holographic fluorescence microscopy. *Opt. Lett.* **1997**, *22*, 1506–1508. [CrossRef] [PubMed]
16. Rosen, J.; Brooker, G. Non-scanning motionless fluorescence three-dimensional holographic microscopy. *Nat. Photon.* **2008**, *2*, 190–195. [CrossRef]
17. Quan, X.; Matoba, O.; Awatsuji, Y. Single-shot incoherent digital holography using a dual-focusing lens with diffraction gratings. *Opt. Lett.* **2017**, *42*, 383–386. [CrossRef]
18. Tahara, T.; Ishii, A.; Ito, T.; Ichihashi, Y.; Oi, R. Single-shot wavelength-multiplexed digital holography for 3D fluorescent microscopy and other imaging modalities. *Appl. Phys. Lett.* **2020**, *117*, 031102. [CrossRef]
19. Tahara, T.; Koujin, T.; Matsuda, A.; Ishii, A.; Ito, T.; Ichihashi, Y.; Oi, R. Incoherent color digital holography with computational coherent superposition for fluorescence imaging [Invited]. *Appl. Opt.* **2021**, *60*, A260–A267. [CrossRef]
20. Tahara, T.; Kozawa, Y.; Ishii, A.; Wakunami, K.; Ichihashi, Y.; Oi, R. Two-step phase-shifting interferometry for self-interference digital holography. *Opt. Lett.* **2021**, *46*, 669–672. [CrossRef]
21. Tahara, T.; Kanno, T.; Arai, Y.; Ozawa, T. Single-shot phase-shifting incoherent digital holography. *J. Opt.* **2017**, *19*, 065705. [CrossRef]
22. Nobukawa, T.; Muroi, T.; Katano, Y.; Kinoshita, N.; Ishii, N. Single-shot phase-shifting incoherent digital holography with multiplexed checkerboard phase gratings. *Opt. Lett.* **2018**, *43*, 1698–1701. [CrossRef] [PubMed]
23. Choi, K.; Joo, K.-I.; Lee, T.-H.; Kim, H.-R.; Yim, J.; Do, H.; Min, S.-W. Compact self-interference incoherent digital holo-graphic camera system with real-time operation. *Opt. Express* **2019**, *27*, 4814–4833. [CrossRef] [PubMed]
24. Liang, D.; Zhang, Q.; Wang, J.; Liu, J. Single-shot Fresnel incoherent digital holography based on geometric phase lens. *J. Mod. Opt.* **2020**, *67*, 92–98. [CrossRef]
25. Imbe, M. Radiometric temperature measurement by incoherent digital holography. *Appl. Opt.* **2019**, *58*, A82–A89. [CrossRef] [PubMed]
26. Nobukawa, T.; Katano, Y.; Goto, M.; Muroi, T.; Hagiwara, K.; Ishii, N. Grating-based in-line geometric-phase-shifting incoherent digital holographic system toward 3D videography. *Opt. Express* **2022**, *30*, 27825–27840. [CrossRef]
27. Tahara, T. Single-shot full-color holography with sunlight. In Proceedings of the OPTICA Digital Holography and 3-D Imaging 2022, Cambridge, UK, 1–4 August 2022. M1A.6.
28. Tahara, T.; Kozawa, Y.; Shimobaba, T. 22 fps motion-picture recording of incoherent holograms with single-shot natural-light full-color digital holography system. In Proceedings of the OSJ Optics and Photonics Japan 2022, Utsunomiya, Japan, 13–16 November 2022. 16pC5. (In Japanese).
29. Vijayakumar, A.; Rosen, J. Spectrum and space resolved 4D imaging by coded aperture correlation holography (COACH) with diffractive objective lens. *Opt. Lett.* **2017**, *42*, 947–950. [CrossRef]
30. Anand, V.; Ng, S.H.; Maksimovic, J.; Linklater, D.; Katkus, T.; Ivanova, E.P.; Judkazis, S. Single shot multispectral multidimensional imaging using chaotic waves. *Sci. Rep.* **2020**, *10*, 13902. [CrossRef]

31. Tahara, T.; Ito, T.; Ichihashi, Y.; Oi, R. Multiwavelength three-dimensional microscopy with spatially incoherent light, based on computational coherent superposition. *Opt. Lett.* **2020**, *45*, 2482–2485. [CrossRef]
32. Tahara, T. Multidimension-multiplexed full-phase-encoding holography. *Opt. Express* **2022**, *30*, 21582–21598. [CrossRef]
33. Tahara, T.; Kozawa, Y.; Matsuda, A.; Oi, R. Quantitative phase imaging with single-path phase-shifting digital holography using a light-emitting diode. *OSA Contin.* **2021**, *4*, 2918–2927. [CrossRef]
34. Tahara, T.; Kozawa, Y.; Oi, R. Single-path single-shot phase-shifting digital holographic microscopy without a laser light source. *Opt. Express* **2022**, *30*, 1182–1194. [CrossRef] [PubMed]
35. Zhu, B.; Ueda, K.-I. Real-time wavefront measurement based on diffraction grating holography. *Opt. Commun.* **2003**, *225*, 1–6. [CrossRef]
36. Millerd, J.; Brock, N.; Hayes, J.; Morris, M.N.; Novak, M.; Wyant, J. Pixelated phase-mask dynamic interferometer. *Proc. SPIE* **2004**, *5531*, 304.
37. Awatsuji, Y.; Sasada, M.; Kubota, T. Parallel quasi-phase-shifting digital holography. *Appl. Phys. Lett.* **2004**, *85*, 1069–1071. [CrossRef]
38. Okamoto, R.; Tahara, T. Precision limit for simultaneous phase and transmittance estimation with phase-shifting interferometry. *Phys. Rev. A* **2021**, *104*, 033521. [CrossRef]
39. Tsuruta, M.; Fukuyama, T.; Tahara, T.; Takaki, Y. Fast image reconstruction technique for parallel phase-shifting digital holography. *Appl. Sci.* **2021**, *11*, 11343. [CrossRef]
40. Shimobaba, T.; Tahara, T.; Hoshi, I.; Shiomi, H.; Wang, F.; Hara, T.; Kakue, T.; Ito, T. Real-valued diffraction calculations for computational holography. *Appl. Opt.* **2022**, *61*, B96–B102. [CrossRef]
41. Hara, T.; Kakue, T.; Shimobaba, T.; Ito, T. Design and implementation of special-purpose computer for incoherent digital holography. *IEEE Access* **2022**, *10*, 76906–76912. [CrossRef]

Proceeding Paper

Imaging Pitfalls and Diagnostic Inhibitions in Various Advanced Head and Neck Imaging Modalities—Diagnostician's Perspective [†]

Durgadevi Boopathi

Department of Oral Medicine and Maxillofacial Radiology, Indira Gandhi Institute of Dental Sciences, Sri Balaji Vidyapeeth (Deemed to be) University, Pondicherry 607402, India; theoralphysician4895@gmail.com or durgaboopathi@gmail.com; Tel.: +91-8838418036

† Presented at the International Conference on "Holography Meets Advanced Manufacturing", Online, 20–22 February 2023.

Abstract: The head and neck is a complex area where imaging plays a major role in not only diagnosis, but also in guided investigations, treatment planning, and, to an extent, guided interventions. This type of imaging ranges from a simple digital orthopantomogram of the jaws and teeth to the complex 3D computed tomography (CT), magnetic resonance imaging (MRI), and cone beam computed tomography (CBCT). Even though the imaging modalities have paved the way for more precise examination and assessment compared to a decade ago; they still require slight renovations in terms of the artifacts and dimensional blurriness invading the diagnosis. This paper sheds light on few of the specific scenarios, such as metal artifacts due to prosthetic crowns, in CT, CBCT, and MRI that greatly hinder the radiological diagnosis and assessment of the extent of lesions, posing critical challenges in surgical planning. With regard to software resolution and available tools, this presentation will cover the restrictions in handling image data, and the processible tools that can be implemented for easy and efficient interpretation and modulation in orthognathic surgery, implant surgery, and excisions of malignancies in the head and neck region.

Keywords: artifacts; imaging software; imaging difficulties; future modifications; head and neck imaging; maxillofacial radiology; oral radiology; imaging

Citation: Boopathi, D. Imaging Pitfalls and Diagnostic Inhibitions in Various Advanced Head and Neck Imaging Modalities—Diagnostician's Perspective. *Eng. Proc.* **2023**, *34*, 1. https://doi.org/10.3390/HMAM2-14354

Academic Editor: Vijayakumar Anand

Published: 19 April 2023

1. Introduction

The advent of advanced imaging modalities has had a great impact and provided advantages in diagnosis and treatment planning in the head and neck region. Being the most complex region, two-dimensional imaging can be used as a preliminary diagnostic order, but complex diagnosis and treatment planning requires advanced imaging.

In the various imaging modalities of the head and neck, we can divide them into those indicated for (i) odontogenic pathology/hard tissue imaging (predominantly cone beam computed tomography) and (ii) non–odontogenic/trauma and malignancy (computed tomography, magnetic resonance imaging, contrast imaging, and PET scans).

There has been an innumerable amount of studies and reviews on the advantages of imaging modalities in the head and neck region. This review is intended to explore the other side—the few disadvantages and diagnostic inhibitions. This paper attempts to bring up solutions and rectifications that can significantly increase the quality of imaging.

2. Cone Beam Computed Tomography

2.1. Imaging Pitfalls

Even though CBCT is used in the case of effective imaging and treatment planning for dental implants, maxillofacial trauma, and assessment of odontogenic pathologies, random artifacts greatly decrease the diagnostic ability. There are many different types of

CT artifacts, including noise, beam hardening, scatter, pseudo enhancement, motion, cone beam, helical, ring, and metal artifacts. Manufacturers minimize beam hardening by using filtration, calibration correction, and beam hardening correction software [1,2].

Consider a case in which the patient presented with chronic left-sided sinusitis. There was a root canal-treated tooth in relation to the left maxillary molar. No clinical signs were evident with regard to teeth. On an intraoral periapical radiograph, obturation seemed to be normal, but revealed well-defined periapical pathology. The extent of the pathology was not assessable, and the periapical pathology was suspected to be the source of the sinusitis infection.

On CBCT, there were beam hardening streaking artifacts (black lines) present in the volume (Figure 1). Upon assessing the axial section at the root end–maxillary sinus floor interface, the streaking artifact had mimicked the destruction of the sinus floor and communication between the sinus and periapical pathology (Figure 2), indicating a radiodiagnosis of odontogenic sinusitis. Upon a surgical exploration with root end surgery, this was proven to be a periapical scar and an intact maxillary sinus floor.

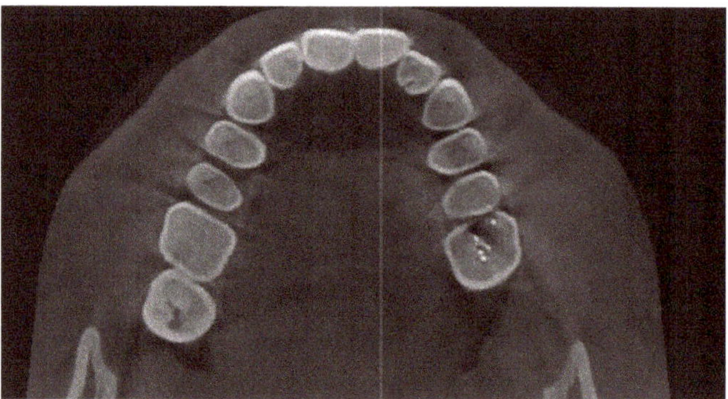

Figure 1. Beam hardening streaking artifacts (black lines) in the axial section.

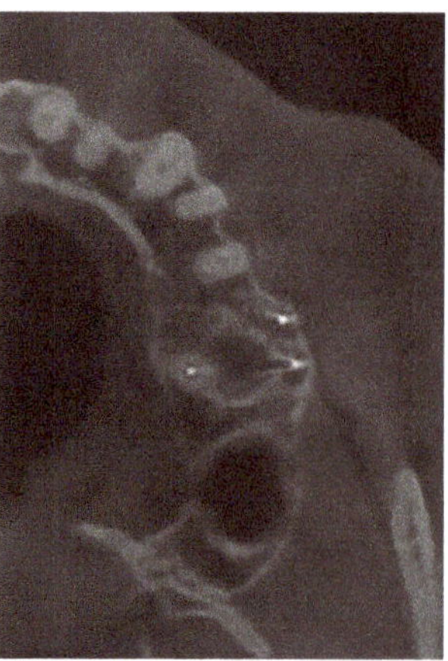

Figure 2. At the root end–maxillary sinus floor interface, the streaking artifact mimicked the destruction of the sinus floor and communication between the sinus and periapical pathology.

The artifacts in CBCT are similar to those in CT, but they are more pronounced in CBCT due to the heterochromatic X-ray beam present in CBCT and the lower mean kilovolt (peak) energy compared with conventional CT.

2.2. Diagnostic Inhibitions

CBCT has very low soft tissue resolution. It is still used for odontogenic/nonodontogenic cysts of the jaws and trauma of the head and neck.

CBCT does not give us differential density values for soft tissues and fluid. The Hounsfield unit and CBCT gray values are highly correlated, which has been proven in multiple research studies [3,4]. Recent studies have attempted to diagnose cysts and soft tissue pathologies through CBCT gray scale values. In a study by Meryem Etöz et al. [5] identifying radicular cysts or apical granulomas using CBCT, a seven-criteria method was used. The criteria included the relationship of lesions with dental roots, periphery of the lesion, shape, darker focus in the center, root resorption, displacement in related teeth, and cortical bone perforation. In addition, the minimum and maximum gray scale values of the lesions were measured and compared. It was concluded that there was no relationship between the histopathological diagnosis of lesions and CBCT gray scale values [5].

This concept can be explained by a case of swelling in the lower jaw for the past 9 months. No associated pain, numbness, or paresthesia were present. Clinically, distal tipping of 32 and mesial tipping of 33 were evident. Upon vitality testing, 42, 41, 31, and 32 were found to be nonvital. A clinical provisional diagnosis of the radicular cyst was given. The aspiration of the lesion revealed a straw-colored fluid. On CBCT, the pathology in the anterior mandible was evident, with buccal cortical bone destruction, and the internal structure revealed thin septa not extending throughout the lesion (Figure 3a,b). This posed a diagnostic dilemma of either a tumor or a cystic lesion. The gray scale value did not provide insight into the internal structure. The histopathological diagnosis was plexiform ameloblastoma, an odontogenic tumor with centric cystic degeneration.

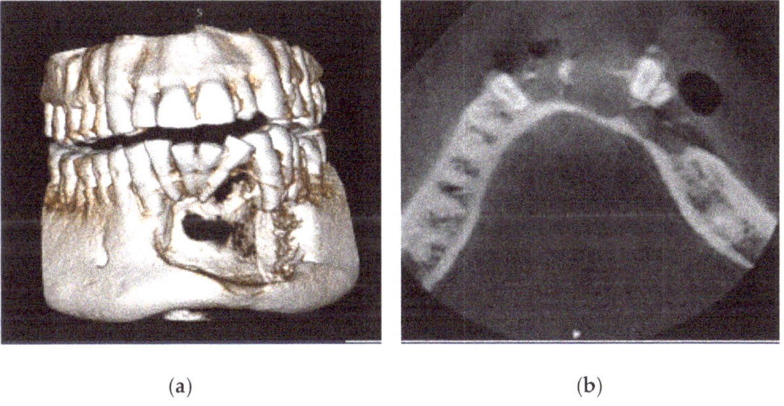

(a) (b)

Figure 3. (**a**) CBCT images of a clinically diagnosed cystic lesion reveal buccal and lingual cortical bone destruction; (**b**) Internal structure revealing thin septa that do not extend throughout the lesion. Gray scale value could not differentiate fluid/soft tissue in this case of plexiform ameloblastoma, thus posing a radiological diagnostic dilemma.

3. Computed Tomography (CT)

Computed tomography has been widely indicated in head and neck trauma, panfacial trauma, and suspected head injuries. CT has been indicated in cases of soft tissue involvement (orbital trauma), pan-facial trauma hindering soft tissue, muscular function, and enophthalmos.

Due to the lack of immobilization of an acute trauma patient, artifacts may be more frequent and nerve canal involvement/muscle involvement may not be assessed.

In a case of zygomatic complex fracture, orbital involvement revealed a linear minimally displaced fracture noted involving the left zygomatic arch, the lateral wall of the left orbit. A comminuted displaced fracture involving the lateral wall of the left maxillary sinus with fracture fragments lying inside was associated with hemosinus (Figures 4 and 5). However, the soft tissue window failed to reveal further details with regard to the pterygoid muscles and orbital volume. Other than the displacement and type of fractures, the extent of orbital involvement also determines surgical treatment in this case. The most common cause of posttraumatic enophthalmos is increased orbital volume [6]. All images were taken using 5th generation mono-energy CT machines with standard exposure parameters of 140 kvp, 200 mA soft tissue with a slice thickness of 0.5–0.6 mm, and bony windows.

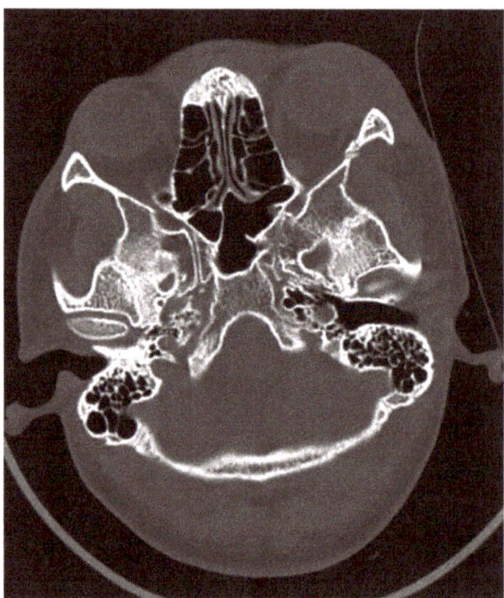

Figure 4. CT axial section of a pan-facial trauma: a comminuted displaced fracture involving the lateral wall of the left maxillary sinus with fracture fragments lying inside, associated with hemosinus.

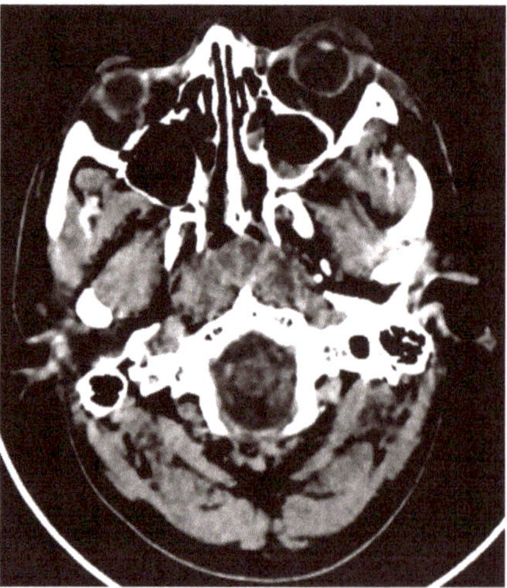

Figure 5. Comminuted displaced fracture involving the lateral wall of the left maxillary sinus with fracture fragments lying inside, associated with hemosinus. The soft tissue window failed to reveal further details with regard to the pterygoid muscles and orbital volume.

One more case of a photon starvation artifact hindering the assessment of the tongue pathology and its extent in a computed tomography image is presented in Figure 6.

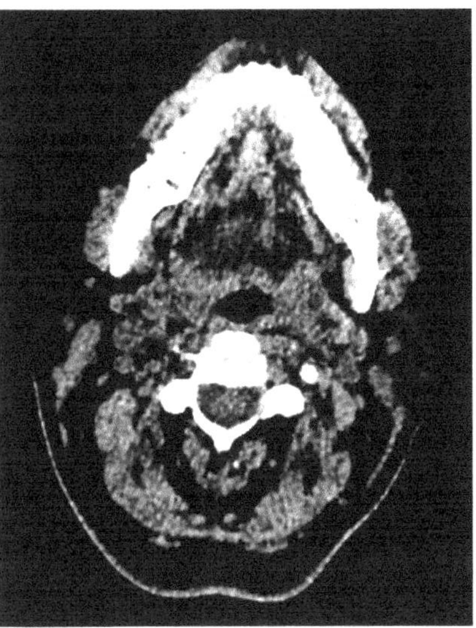

Figure 6. Photon starvation artifact hindering the assessment of the tongue pathology and its extent in a computed tomography image.

This photon starvation, or quantum mottle, greatly hindered the diagnostic quality of the image. This results from insufficient X-ray photons reaching the sensor from the patient. Even though techniques such as tube current modulation (TCM) and vendor-specific and iterative reconstruction algorithms can reduce this effect, these types of artifacts still represent considerable hindrances to diagnosis and assessment.

Streak and windmill artifacts greatly affect the scan images in CT. Windmill artifacts occur due to under-sampling along the Z axis, which usually occurs in the clavicle region and the base of the skull region, where drastic anatomical changes and differential Hounsfield units are present. A black streak artifact of the beam hardening effect greatly disrupting the diagnostic quality can be seen in Figure 7.

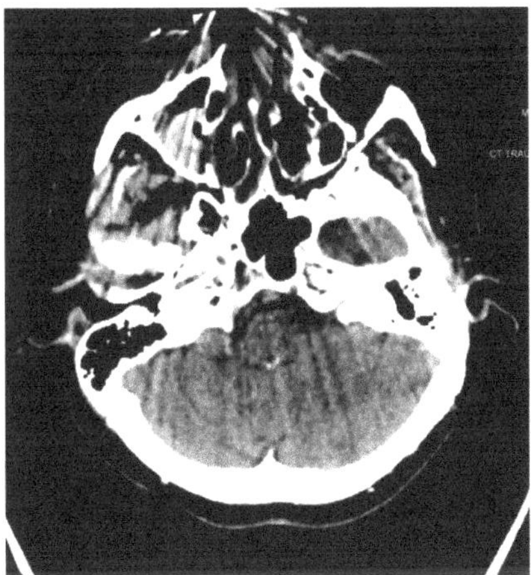

Figure 7. Black streak artifact of beam hardening effect evident in an axial section CT image. These are usually present at the base of the skull region, where drastic anatomical changes and differential Hounsfield units are present.

4. Magnetic Resonance Imaging (MRI)

Magnetic resonance imaging is used in head and neck imaging predominantly for ruling out craniofacial neuralgias, soft tissue extension of pathologies and malignancies of the head and neck, diagnosis, and treatment planning.

The most common artifact present in MRI is motion blurriness (due to long scan time) and metal artifacts.

Motion blurriness happens due to patient motion, swallowing, and even breathing in the case of the head and neck region. The degree of blurriness could fall into a range of minimal to extensive blur. A pediatric MRI image with severe motion blurs and distortion of the image is presented in the sagittal section (Figure 8). This kind of artifact requires a re-scan.

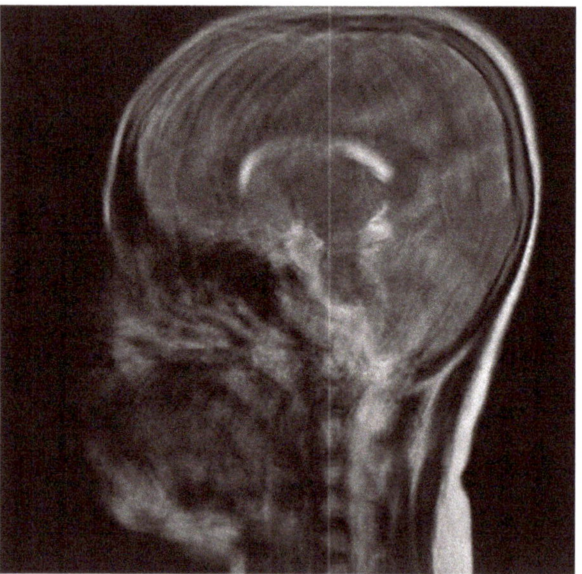

Figure 8. MRI image of a pediatric patient taken for neuroimaging purposes. Severe motion blurs and distortion of the image are evident.

Metal artifacts manifest in multiple forms—a complete signal loss may be observed around the metallic object, or a rim of high signal intensity may be present around the metallic devices. Metal-induced artifacts are still a challenging aspect of imaging. Many technical methods and minor alterations have been proposed to reduce the presence of metal-induced artifacts in MRI, such as view-angle tilting method, MAVRIC (multi-acquisition variable resonance image combination), and SEMAC (slice-encoding for metal artifact correction) can be used in the plane and through-slice displacements, but at a cost of increased scan time, further increasing the inconvenience [7–9].

A similar example is presented in the T2 axial section (Figure 9). The porcelain fused to metal fixed prosthesis placed in the mandibular anterior teeth caused a no-signal, presenting anterior artifact extending to the level of the maxillary teeth. This is a particularly inconvenient problem since the MRI imaging was performed to check for any malignant spread in a recurring cancer patient. Although different technical alterations gave us a clue about the region in one plane, it was not sufficient.

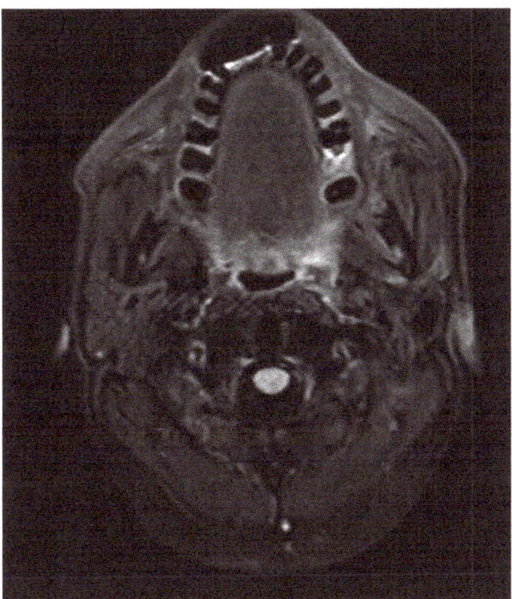

Figure 9. T2 weighted image—axial section showing loss of signal in anterior mandible and maxilla due to metal fused crown in mandibular anterior. This obscures the purpose of the MRI, which was performed to identify any malignant spread in a recurring cancer patient.

Positron Emission Tomography

PET-CT (Positron Emission Tomography—Computed Tomography) is used to identify, locate, and assess the anatomy and function of hypermetabolic tissue. It uses 18-FDG (fluorodeoxyglucose) as a medium. The signals are then read by the system. 18-FDG PET-CT is one of the most accurate methods for diagnosing primary HNC lesions. The lesion cannot be accessed via visual inspection, with or without concurrent vision-guided biopsy [10]. The specificity of PET-CT for identifying inspection-occult oropharyngeal SCC has been measured at over 90%. According to the NCCN (National Comprehensive Cancer Network), PET-CT is indicated for initial diagnosis for the following types of physical examination for occult HNC (Head and Neck Cancers): oral cavity, supraglottic larynx, ethmoid sinus, glottic larynx, oropharynx, and maxillary sinus [11,12].

It is widely used as a recall–review imaging modality for post-treatment follow-up cases. The false positive possibility of PET-CT poses a major diagnostic inhibition and a major pitfall. This is the reason why PET-CT has been avoided for post-radiation and therapy patients for at least up to 8–12 weeks (to avoid post-inflammatory effects) [13].

Even then, PET-CT might show false positive results in a review patient who poses diagnostic confusion. One such case is presented here: A 58-year-old male patient was diagnosed and treated for salivary gland ductal carcinoma by surgery and radiotherapy one year prior to this event. In the review CT, brain metastasis was evident. A PET-CT was advised immediately, and PET-CT revealed metastasis to the lung nodules as well.

In addition to the signals in the metastatic nodules, a well-defined positive signal was evident in the submental region (Figure 10a,b). Upon clinical correlation, there was no evidence of a sign of metastasis in the submental region. This might be confused with a metastatic involvement, thus representing a pitfall.

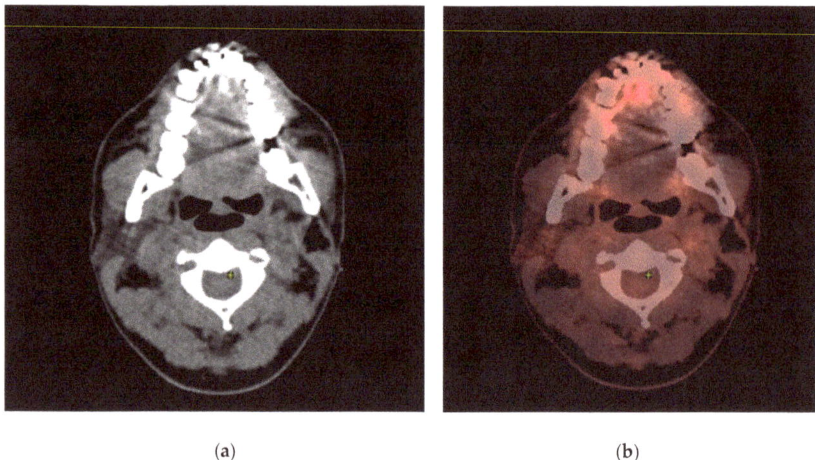

(a)   (b)

Figure 10. (**a**) CT reveals hyperdense area in the submental region; (**b**) PET-CT reveals a well-defined positive signal was evident in the submental region, which was a false positive finding.

This review briefly compiles the artifacts and diagnostic inhibitions/restrictions in head and neck imaging. Little software has been developed to reduce metal-induced artifacts, but its usage is limited and not accessible. Metal artifact reduction software (MARs) has been tested in phantom experiments with dental implants using mono-energy (MonoE) CT images, and it was found that a combination of MonoE and MARs reconstruction was the best method for reducing metal artifacts. This study had some major limitations [14].

5. Conclusions

The advanced imaging in the head and neck region has greatly lifted the quality and efficiency of the diagnosis and treatment planning of pathologies. As new techniques and software are being introduced, their availability and accessibility are considerably lower. Even after advancements in imaging, these modalities have their pitfalls and diagnostic inhibitions. This paper attempts to address these from a maxillofacial physician and radiodiagnosis point of view, which can be used by physicists for the development of more accessible software and technical alterations to reduce these issues.

Funding: This research received no external funding.

Institutional Review Board Statement: Not applicable.

Informed Consent Statement: Patient consent was waived since the paper does not reveal any patient identification and uses only radiographic images from archives.

Data Availability Statement: Not applicable.

Acknowledgments: I deeply acknowledge and thank my friend and colleague Sai Kishore, Department of Nuclear Medicine, JIPMER, Puducherry for his contribution and sharing of the PET-CT images.

Conflicts of Interest: The authors declare no conflict of interest.

References

1. Barrett, J.F.; Keat, N. Artifacts in CT: Recognition and Avoidance. *RadioGraphics* **2004**, *24*, 1679–1691. [CrossRef] [PubMed]
2. Dwivedi, N.; Nagarajappa, A.K.; Tiwari, R. Artifacts: The downturn of CBCT image. *J. Int. Soc. Prev. Community Dent.* **2015**, *5*, 440–445. [CrossRef] [PubMed]
3. Razi, T.; Niknami, M.; Ghazani, F.A. Relationship between Hounsfield Unit in CT Scan and Gray Scale in CBCT. *J. Dent. Res. Dent. Clin. Dent. Prospect.* **2014**, *8*, 107–110. [CrossRef]

4.   Mah, P.; Reeves, T.E.; McDavid, W.D. Deriving Hounsfield units using grey levels in cone beam computed tomography. *Dentomaxillofacial Radiol.* **2010**, *39*, 323–335. [CrossRef] [PubMed]

5.   Etöz, M.; Amuk, M.; Avcı, F.; Yabacı, A. Investigation of the effectiveness of CBCT and gray scale values in the differential diagnosis of apical cysts and granulomas. *Oral Radiol.* **2021**, *37*, 109–117. [CrossRef] [PubMed]

6.   Winegar, B.A.; Murillo, H.; Tantiwongkosi, B. Spectrum of critical imaging findings in complex facial skeletal trauma. *RadioGraphics* **2013**, *33*, 3–19. [CrossRef]

7.   Hargreaves, B.A.; Worters, P.W.; Pauly, K.B.; Pauly, J.M.; Koch, K.M.; Gold, G.E. Metal-Induced Artifacts in MRI. *Am. J. Roentgenol.* **2011**, *197*, 547–555. [CrossRef]

8.   Koch, K.M.; Brau, A.C.; Chen, W.; Gold, G.E.; Hargreaves, B.A.; Koff, M.; McKinnon, G.C.; Potter, H.G.; King, K.F. Imaging near metal with a MAVRIC-SEMAC hybrid. *Magn. Reson. Med.* **2010**, *65*, 71–82. [CrossRef] [PubMed]

9.   Lu, W.; Pauly, K.B.; Gold, G.E.; Pauly, J.M.; Hargreaves, B.A. SEMAC: Slice encoding for metal artifact correction in MRI. *Magn. Reson. Med.* **2009**, *62*, 66–76. [CrossRef] [PubMed]

10. Boellaard, R.; Delgado-Bolton, R.; Oyen, W.J.G.; Giammarile, F.; Tatsch, K.; Eschner, W.; Verzijlbergen, F.J.; Barrington, S.F.; Pike, L.C.; Weber, W.A.; et al. FDG PET/CT: EANM procedure guidelines for tumour imaging: Version 2.0. *Eur. J. Nucl. Med. Mol. Imaging* **2015**, *42*, 328–354. [CrossRef] [PubMed]

11. Padhani, A.R. Imaging in the evaluation of cancer. In *Recommendations for Cross-Sectional Imaging in Cancer Management*; Nicholson, T., Ed.; The Royal College of Radiologists: London, UK, 2014.

12. Kawabe, J.; Higashiyama, S.; Yoshida, A.; Kotani, K.; Shiomi, S. The role of FDG PET-CT in the therapeutic evaluation for HNSCC patients. *Jpn. J. Radiol.* **2012**, *30*, 463–470. [CrossRef] [PubMed]

13. Agrawal, A.; Rangarajan, V. Appropriateness criteria of FDG PET/CT in oncology. *Indian J. Radiol. Imaging* **2015**, *25*, 88–101. [CrossRef] [PubMed]

14. Sun, X.; Zhao, Q.; Sun, P.; Yao, Z.; Wang, R. Metal artifact reduction using mono-energy images combined with metal artifact reduction software in spectral computed tomography: A study on phantoms. *Quant. Imaging Med. Surg.* **2020**, *10*, 1515–1525. [CrossRef] [PubMed]

Proceeding Paper

Enhancing Phase Measurement by a Factor of Two in the Stokes Correlation [†]

Amit Yadav [1],*, Tushar Sarkar [1], Takamasa Suzuki [2] and Rakesh Kumar Singh [1]

[1] Laboratory of Information Photonics and Optical Metrology, Department of Physics,
   Indian Institute of Technology (Banaras Hindu University), Varanasi 221005, India;
   tusharsarkar.sarkar@gmail.com (T.S.); krakeshsingh.phy@iitbhu.ac.in (R.K.S.)
[2] Department of Electrical and Electronic Engineering, Niigata University, Niigata 950-2181, Japan;
   takamasa@eng.niigata-u.ac.jp
* Correspondence: yadavamitupac@gmail.com
† Presented at the International Conference on "Holography Meets Advanced Manufacturing", Online,
   20–22 February 2023.

Abstract: Phase loss is a typical problem in the optical domain, and optical detectors only measure the amplitude distribution of a signal without its phase. However, an optimal phase is desired in a variety of practical applications, such as optical metrology, nondestructive testing, and quantitative microscopy. Several methods have been proposed to quantitatively measure phase, among which interferometry is one of the most commonly used. An intensity interferometer has also been used to recover phase and enhance the phase difference measurement via the intensity correlation. In this paper, we present and examine another technique based on the Stokes correlation for enhancing phase measurement by a factor of two. The enhancement in phase measurement is accomplished through an evaluation of the correlation between two points of Stokes fluctuations of randomly scattered light and by recovering the enhanced phase of the object by using three-step phase shifting along with the Stokes correlations. This technique is expected to be useful for imaging and the experimental measurement of the phase of a weak signal.

Keywords: vortex beam; speckle; coherence; phase imaging

Citation: Yadav, A.; Sarkar, T.; Suzuki, T.; Singh, R.K. Enhancing Phase Measurement by a Factor of Two in the Stokes Correlation. *Eng. Proc.* **2023**, *34*, 4. https://doi.org/10.3390/HMAM2-14273

Academic Editor: Vijayakumar Anand

Published: 23 May 2023

1. Introduction

Among the most active areas of research in contemporary physics is optical metrology [1], which aims to increase measurement accuracy. This encourages the study of novel phenomena and the advancement of fundamental research. Phase measurement is important in metrology for measuring length, speed, rotation vibration, etc. Phase loss is a common problem in the optical domain, and optical detectors only measure the amplitude distribution of a signal without its phase. Several methods have been proposed to quantitatively measure phase [2], among which interferometry is one of the most commonly used. However, the majority of these methods are only capable of detecting the phase profile of a phase object in free space or homogeneous media. Intensity interferometers have also been used to recover the phase and enhance the phase difference measurement according to the intensity correlation [3,4]. Measuring weak phase information is a common problem in this regard. It has been suggested that the enhancement of the phase during measurement might increase precision and phase resolution.

In this paper, we present and examine a highly stable non-interferometric technique for recovering and enhancing phase measurement by a factor of two through scattering media. This is realized using a two-point Stokes correlation along with three-step phase shifting [5]. Our method employs the complex polarization correlation function (CPCF), which provides a complex Fourier coefficient and enhanced phase measurement. The complex Fourier coefficient retrieves information on the complex amplitude and uses it to extract enhanced phase information. The pilot-assisted strategy, which loads a phase object into one of the orthogonal polarization states, is employed to design a compact, highly stable, and robust noninterferometric setup. The two stokes parameters S_2 and S_3 are used as the theoretical basis of our method.

2. Theory

Consider a polarized, coherent light source that is traveling along the z-axis and has two orthogonal polarization states: x and y. A phase object is loaded on one polarization state of the beam and the other is left unloaded. At the transverse plane $z = 0$, the complex field of coherent-polarized light is represented as follows:

$$E(\hat{r}) = E_x(\hat{r})\hat{e}_x + E_y(\hat{r})\,\hat{e}_y \tag{1}$$

where $E_x(\hat{r})$ *and* $E_y(\hat{r})$ represent the x and y polarization components of the beam, respectively. A light beam (given in Equation (1)) passes through the random scattering media and travels to the detection plane, which is located at any random distance z as determined by the Fresnel diffraction formula. At the arbitrary distance z, the scattered field is represented as follows:

$$E_d(r) = \int E_d(\hat{r})e^{i\delta(\hat{r})}g(r,\hat{r})d\hat{r}, \quad d = (x,\,y) \tag{2}$$

where $g(r,\hat{r}) = \frac{-ik}{2\pi z} \exp\{\frac{ik}{2}[\frac{(r-\hat{r})^2}{z}]\}$ is a propagation kernel, and k is a wavenumber. The position vector at the detector plane and source are denoted by r and $\hat{r}$, respectively, and the diffuser introduces a random phase $(\hat{r})$. Pauli spin matrices are used to define the Stokes parameter of the scattered field as follows:

$$S_n = E^T(\hat{r})\sigma^n E * (\hat{r}), \quad n \in (0,\dots,3) \tag{3}$$

σ^0 is the identity matrix, and σ^1, σ^2, and σ^3 are the Pauli spin matrices of a 2×2 order. The Stokes fluctuation around the mean value of the SPs is as follows:

$$\Delta S_n(r) = S_n(r) - < S_n(r) > \tag{4}$$

where brackets $(< >)$ represent the ensemble average. Let us consider that the behavior of the random light beam corresponds to Gaussian statistics; accordingly, the Gaussian moment theorem is used to illustrate the SPs' fluctuations as follows:

$$C_{pq}(r_1, r_2) = < \Delta S_p(r_1)\Delta S_q(r_2) > \quad \text{where} \ p, q \in (0,\,\dots,\,3) \tag{5}$$

From the above equation, $C_{22}(r_1, r_2)$ and $C_{33}(r_1, r_2)$ are calculated. The real parts of the CPCF are obtained by subtracting $C_{33}(r_1, r_2)$ from $C_{22}(r_1, r_2)$ as follows:

$$C_{Re}(r_1, r_2) = C_{22}(r_1, r_2) - C_{33}(r_1, r_2) \propto Re\left[W_{xy}(r_1, r_2)W_{yx}^*(r_1, r_2)\right] \tag{6}$$

where $W_{xy}(r_1, r_2)W_{yx}^*(r_1, r_2) = < E_x^*(r_1)E_y(r_2) > [< E_y^*(r_1)E_x(r_2) >]^*$.

Now, we use Equation (6) in the development of a phase recovery and enhancement method through scattering media. Here, a phase object named vortex beam, i.e., $E_x(\hat{r}) = A\exp(il\varphi)$ with a topological charge l and an azimuthal index φ, is loaded into the x-polarized state, while the y-polarized state is reserved as a reference beam, i.e., $E_y(\hat{r}) = B$, where A and B represent the amplitude distribution of the vortex beam and the plane beam, respectively. The values of $W_{xy}(r_1, r_2)$ and $W_{yx}^*(r_1, r_2)$ are substituted into Equation (6). Therefore, Equation (6) transforms [6] into

$$C_{Re}(r_1, r_2) \propto Re\left[< E_x^*(r_1)E_y(r_2) > \left\{< E_y^*(r_1)E_x(r_2) >\right\}^*\right] \tag{7}$$

Consequently, Equation (7) is combined with the three-step phase-shifting method [7] to obtain CPCF, which is given below

$$C(\Delta r) = 2C_{Re}^{0}(\Delta r) - C_{Re}^{2\pi/3}(\Delta r) - C_{Re}^{4\pi/3}(\Delta r) + \sqrt{3}i\left[C_{Re}^{2\pi/3}(\Delta r) - C_{Re}^{4\pi/3}(\Delta r)\right] \qquad (8)$$

where $C_{Re}^{4\pi/3}(\Delta r)$, $C_{Re}^{2\pi/3}(\Delta r)$, and $C_{Re}^{0}(\Delta r)$ indicate the real components of the CPCF with a phase shift of $\frac{4\pi}{3}$, $\frac{2\pi}{3}$, and 0, respectively. $C(\Delta r)$ is our required quantity and is utilized to recover and enhance the phase measurement of the phase object.

A schematic diagram of our experiment is shown in Figure 1 given below.

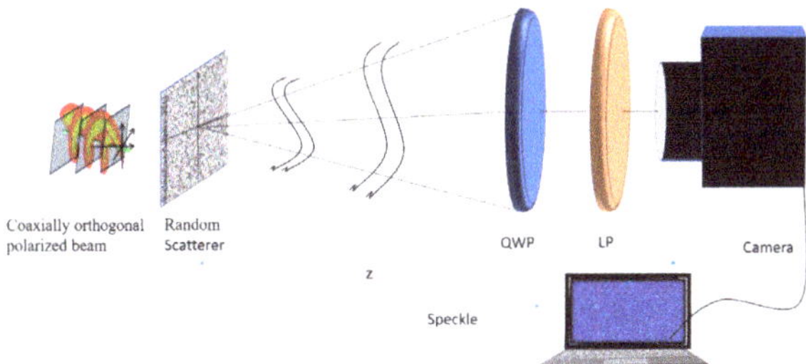

Figure 1. Schematic representation of proposed method. QWP—quarter wave plate; LP—linear polarizer. The CCD records intensity speckle patterns at the observation plane. These speckle patterns are used to determine two SPs.

3. Result and Discussion

We load a vortex beam with a topological charge of $l = -1$ in x polarization component of the beam and use y polarization as a guide. The vortex beam is considered a phase object with which to test the enhancement in the phase measurement. Around the singularity, the vortex beam's phase change is on the order of $2l\pi$. For the topological charge $l = -1$, the phase variation is one order of magnitude greater than 2π around the singularity. However, reconstructed results from the polarization correlations based on Equation (7) show a topological charge of $l = -2$ instead of $l = -1$. These results present a two-fold enhancement in phase measurement. The simulation and experimental results of our proposed method are shown in Figure 2.

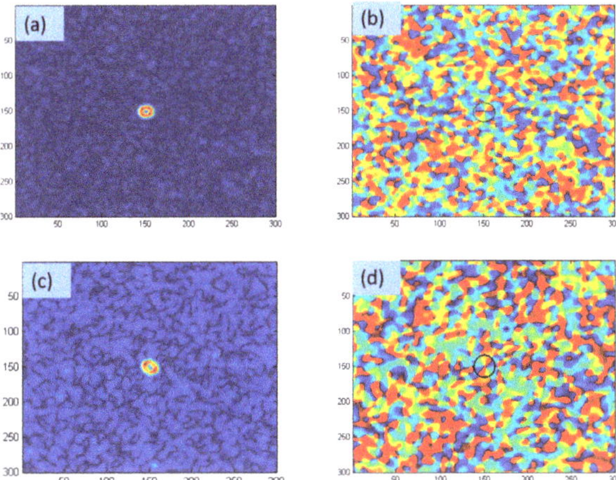

Figure 2. Simulation results: (**a**) amplitude distribution and (**b**) corresponding phase distribution for vortex beam with a charge of $l = -1$; Experimental results: (**c**) amplitude distribution and (**d**) corresponding phase distribution for the vortex beam with $l = -1$.

4. Conclusions

We have presented a method for enhancing phase measurement by using higher-order stokes fluctuation correlations. This method is expected to be helpful in optical metrology and the measurement of weak phase information. The proposed method's viability was assessed via numerical simulation, which was followed by an experimental demonstration to gain enhanced phase information.

Author Contributions: Conceived the idea, prepared manuscript, and methodology, A.Y.; manuscript preparation, methodology, and experimental design, T.S. (Tushar Sarkar); provided advice, reviewed the manuscript, supervised the research, and edited the work, T.S. (Takamasa Suzuki); involved in supervision, formulation of research goals, funding acquisition, and reviewing and editing, R.K.S. All authors have read and agreed to the published version of the manuscript.

Funding: This work is supported by the Science and Engineering Research Board (SERB) India CORE/2019/000026.

Institutional Review Board Statement: Not applicable.

Informed Consent Statement: Not applicable.

Data Availability Statement: The data used in this paper are not currently publicly accessible; however, they are available from the authors upon justifiable request.

Acknowledgments: Amit Yadav acknowledges University Grant Commission, India, for its financial support through the Junior Research Fellowship.

Conflicts of Interest: The authors declare no conflict of interest.

References

1. De Groot, P.J. A review of selected topics in interferometric optical metrology. *Rep. Prog. Phys.* **2021**, *82*, 056101. [CrossRef] [PubMed]
2. Creath, K. V Phase-Measurement Interferometry Techniques. In *Progress in Optics*; Wolf, E., Ed.; Elsevier: Amsterdam, The Netherlands, 1988; Volume 26, pp. 349–393.
3. Shirai, T. Phase difference enhancement with classical intensity interferometry. *Opt. Commun.* **2016**, *380*, 239–244. [CrossRef]
4. Singh, R.K.; Vinu, R.V.; Sharma, A.M. Recovery of complex valued objects from two-point intensity correlation measurement. *Appl. Phys. Lett.* **2014**, *104*, 111108. [CrossRef]
5. Kuebel, D.; Visser, T.D. Generalized Hanbury Brown-Twiss effect for Stokes parameters. *J. Opt. Soc. Am.* **2019**, *36*, 362. [CrossRef] [PubMed]
6. Sarkar, T.; Parvin, R.; Brundavanam, M.M.; Kumar Singh, R. Higher-order Stokes-parameter correlation to restore the twisted wave front propagating through a scattering medium. *Phys. Rev. A* **2021**, *104*, 013525. [CrossRef]
7. Huang, P.S.; Zhang, S. Fast three-step phase-shifting algorithm. *Appl. Opt.* **2006**, *45*, 5086–5091. [CrossRef] [PubMed]

Proceeding Paper

Holography with Incoherent Light [†]

Akanksha Gautam [1,*], Athira T S [2], Dinesh N. Naik [2], C. S. Narayanmurthy [2], Rajeev Singh [1] and Rakesh Kumar Singh [1]

1   Laboratory of Information Photonics and Optical Metrology, Department of Physics,
    Indian Institute of Technology (Banaras Hindu University), Varanasi 221005, Uttar Pradesh, India;
    rajeevs.phy@iitbhu.ac.in (R.S.); krakeshsingh.phy@iitbhu.ac.in (R.K.S.)
2   Applied and Adaptive Optics Laboratory, Department of Physics, Indian Institute of Space Science
    and Technology, Thiruvananthapuram 695547, Kerala, India; athira.18@res.iist.ac.in (A.T.S.);
    dineshnnaik@iist.ac.in (D.N.N.); murthy@iist.ac.in (C.S.N.)
*   Correspondence: akankshagautam.rs.phy19@itbhu.ac.in
†   Presented at the International Conference on "Holography Meets Advanced Manufacturing", Online,
    20–22 February 2023.

Abstract: Conventional digital holography uses the technique of combining two coherent light fields and the numerical reconstruction of the recorded hologram leads to the object amplitude and phase information. Despite significant developments in the DH with coherent light, complex field imaging with arbitrary coherent sources is also desired for various reasons. Here, we present a possible experimental approach for holography with incoherent light. In the case of incoherent light, the complex spatial coherence function is a measurable quantity and the incoherent object holograms are recorded as the coherence function. Thus, to record complex spatial coherence a square Sagnac radial shearing interferometer is designed with the phase-shifting approach. The five-step phase-shifting method helps to measure the fringe visibility and the corresponding phase, which jointly represents the complex coherence function. The inverse Fourier transform of the complex coherence function helps to retrieve the object information.

Keywords: digital holography; interference; coherence; phase-shifting

1. Introduction

Digital holography (DH) is a technique of interference of two waves both spatially and temporally coherent [1]. The recorded interference pattern contains both amplitude and phase information which can be further reconstructed by digitally processing the recorded hologram [2,3]. The coherent source is utilized to achieve the interference in the DH, but the coherent imaging system suffers from speckle noise and edge effects [4,5], whereas the incoherent imaging systems both spatially and temporally incoherent, such as broadband or a light emitting diode (LED), do not suffer from speckle noise and also cost-effective. Therefore, over coherent imaging systems incoherent imaging systems are preferred for certain applications.

The interference of incoherent light reflected or emitted from an object result in incoherent digital holograms. Many known methods of recording incoherent holograms are based on the self-interference principle which uses the property that each incoherent source point is acting as an independent scatterer and is self-spatially coherent; hence, it can generate an interference pattern with light coming from its mirror imaged point [6–8]. These techniques record the incoherent hologram in the form of intensity patterns. Another way to record an incoherent hologram is through a two-point spatial complex coherence function [9,10]. These methods are based on the van Cittert–Zernike (VCZ) theorem which connects the far-field complex coherence function with the incoherent source intensity distribution.

In this paper, we present an experimental approach to recording an incoherent object hologram based on the VCZ theorem and present some initial results. In the first part of the

Citation: Gautam, A.; T S, A.; Naik, D.N.; Narayanmurthy, C.S.; Singh, R.; Singh, R.K. Holography with Incoherent Light. *Eng. Proc.* **2023**, *34*, 5. https://doi.org/10.3390/HMAM2-14111

Academic Editor: Vijayakumar Anand

Published: 4 March 2023

set up the object information is recorded as a two-point complex spatial coherence function and in the later part a square Sagnac radial shearing interferometer with a five-step phase-shifting technique is designed to obtain the complex coherence function. The digital inverse Fourier transform of complex coherence function helps to retrieve the object information.

2. Principle

The principle of coherence holography is based on the existing similarity between the VCZ theorem and the diffraction integral. The VCZ theorem connects the coherence function of the incoherent source with its intensity distribution through the Fourier transform relation. Consider a resolution target (number '5') incoherently illuminated by a yellow LED as shown in Figure 1, having spectral scattering density $\eta(\tilde{r}; \lambda)$ where λ is the mean wavelength of the LED and $\tilde{r}$ is the transverse coordinate, respectively. As the source is incoherent each point of LED acts as a random scatterer having instantaneous phase $\phi(\tilde{r}, t)$. The spectral component of the field is given by $A(\tilde{r}; \lambda) = \sqrt{\eta(\tilde{r}; \lambda)} \exp[i\phi(\tilde{r}, t)]$ corresponding to λ. The field is Fourier transformed using a lens L having a focal length f and at the Fourier plane, a single realization of the field is represented as

$$E(r) = \iint A(\tilde{r}; \lambda) \exp\left(-\frac{i2\pi}{\lambda f}(r.\tilde{r})\right) d\tilde{r} \tag{1}$$

where r is the coordinate of Fourier plane.

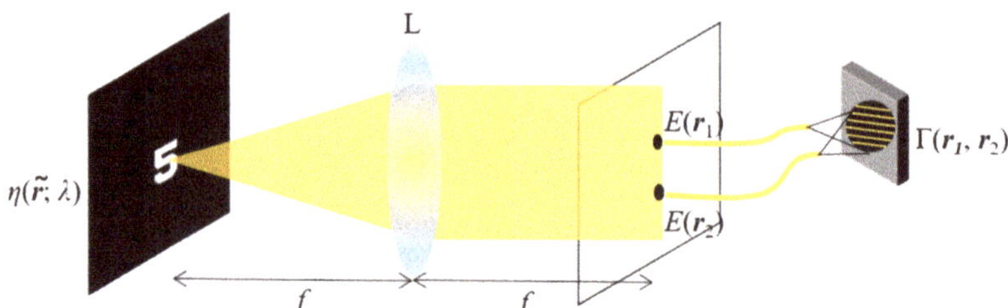

Figure 1. Geometry for recording information of spatial coherence function.

The coherence function at the Fourier plane is given by

$$\Gamma(r_1, r_2) = E^*(r_1)E(r_2) \tag{2}$$

where the asterisk represents complex conjugate and angle bracket denotes the ensemble average.

From Equation (1), we can write

$$\Gamma(r_1, r_2) = \frac{\kappa}{\lambda^2 f^2} \iint \sqrt{\eta(\tilde{r}_1; \lambda)\eta(\tilde{r}_2; \lambda)} \langle \exp[i(\phi(\tilde{r}_2, t) - \phi(\tilde{r}_1, t)]\rangle \exp\left[-i\frac{2\pi}{\lambda f}(r_2.\tilde{r}_2 - r_1.\tilde{r}_1)\right] d\tilde{r}_1 d\tilde{r}_2 \tag{3}$$

where κ is a constant physical quantity having dimensions of length square. As the LED source is spatially incoherent any two points of the source are mutually uncorrelated and hence represented by a two-dimensional (2-D) Dirac–delta function

$$\langle \exp[i(\phi(\tilde{r}_2, t) - \phi(\tilde{r}_1, t)]\rangle = \delta(\tilde{r}_2 - \tilde{r}_1)$$

Equation (3) is now reduced to

$$\Gamma(r_1, r_2) = \frac{\kappa}{\lambda^2 f^2}\left\{\int \eta(\tilde{r}; \lambda) \exp\left[-i\frac{2\pi}{\lambda f}\tilde{r}.(r_2 - r_1)\right] d\tilde{r}\right\} d\lambda \tag{4}$$

In the above equation an integration is also performed over the wavelength range λ as the source is non-monochromatic and have a finite spectral bandwidth. The integral inside the curly bracket represents van Cittert–Zernike theorem.

3. Experiment and Results

Figure 2 shows the experimental setup to record the spatial coherence function. The first part of the setup shows the recording of object information as explained in the previous section. A polarizer P1 is kept at 45° just after the LED source to make the unpolarized light coming from the LED polarized. The field is Fourier transformed using a lens L1 kept at its focal length $f_1 = 60$ mm from the target, number '5'. The black dotted line shows the Fourier plane where the spatial coherence function represented by Equation (4) is present. In the later part, a square Sagnac radial shearing interferometer with a telescopic lens system L2 ($f_2 = 120$ mm) and L3 ($f_3 = 125$ mm) is designed to measure the coherence function. The incoming field is divided by a polarizing beam splitter with orthogonal polarization states in two parts and between the two oppositely counter-propagating beams one gets magnified with magnification $\alpha = f_3/f_2$ and the other gets demagnified with magnification $\alpha^{-1} = f_2/f_3$. Finally, at the output plane shown by blue dotted lines, we get radially sheared copies of the two fields. The output plane parameters are now scaled as $r_1 = \alpha^{-1}r$ and $r_2 = \alpha r$ and the coherence function; $\Gamma(r_1, r_2)$ is now mapped as $\Gamma(\alpha^{-1}r, \alpha r)$. The output plane does not fit the CMOS camera area (Thorlabs DCC3240M, imaging area 6.78 mm × 5.43 mm with pixel size 5.3 µm); thus, an imaging lens L4 with focal length $f_4 = 150$ mm is used to demagnify the image such that it fits the camera aperture. At the output, just before the camera, a quarter-wave plate (QWP) is kept at 45° from its fast axis to convert the two linear orthogonal polarization states into right circular and left circular polarization states, respectively. Later, a polarizer P2 is kept and rotated by angle θ such that it introduces a phase shift 2θ between the two incoming beams. Five off-axis interferograms with phase shift 0, $\pi/2$, π, $3\pi/2$, and 2π are recorded by giving a tilt using mirror M2 between the two counter-propagating beams. The fringe contrast γ and corresponding phase $\phi\prime$ can be calculated as mentioned in references [11,12], respectively. Thus, the complex spatial coherence function is built as $\Gamma(\alpha^{-1}r, \alpha r) = \gamma \exp(i\phi\prime)$.

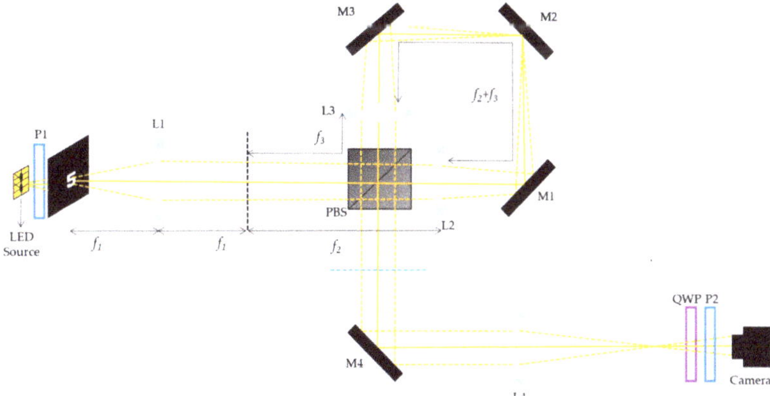

Figure 2. Experimental set up square Sagnac radial shearing interferometer: P: Polarizer, L: Lens, BS: Beam Splitter, M: Mirror, QWP: Quarter Wave Plate.

Now, to reconstruct the object intensity distribution the spatial coherence function is inverse Fourier transformed.

$$\tilde{\eta}(\tilde{r}; \lambda) = \left(\alpha - \alpha^{-1}\right)^2 \int \Gamma\left(\alpha^{-1}r, \alpha r\right) \exp\left[i\frac{2\pi}{\lambda f}\left(\alpha - \alpha^{-1}\right)r.\tilde{r}\right] dr \tag{5}$$

Figure 3 represents the digitally constructed (a) fringe visibility and (b) corresponding phase using the five recorded interferograms and (c) the reconstructed object intensity distribution showing the information of number '5'.

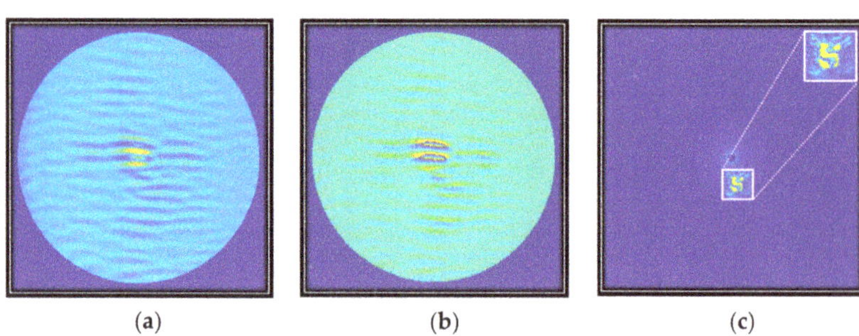

Figure 3. Digitally constructed coherence function (**a**) fringe visibility, (**b**) corresponding phase, and (**c**) reconstructed object intensity distribution.

4. Conclusions

We have presented an experimental method using shearing for holography with an incoherent source. The object information is recorded in the form of a complex spatial coherence function based on the principle of van Cittert–Zernike theorem and later it is analyzed using a Sagnac radial shearing interferometer with the five-phase shifting algorithm. The object information is computationally acquired on inverse Fourier transform.

Author Contributions: A.G.: Conceived the idea, Investigation, Writing manuscript, Methodology, Experimental Design. A.T.S.: Experimental Design. D.N.N.: Ideas, Revision and Editing, Supervision. C.S.N.: Editing and revision of the manuscript. R.S.: Editing and revision of the manuscript. R.K.S.: Ideas, Formulation of research goals, Revision and Editing, Funding acquisition, Supervision. All authors have read and agreed to the published version of the manuscript.

Funding: This work is supported by the Board of Research in Nuclear Sciences (BRNS—Grant No. 58/14/04/2021-BRNS/37092).

Institutional Review Board Statement: Not applicable.

Informed Consent Statement: Not applicable.

Data Availability Statement: Data underlying the results presented in this paper are not publicly available at this time but may be obtained from the authors upon reasonable request.

Acknowledgments: Akanksha Gautam would like to acknowledge support from DST-INSPIRE (IF180930).

Conflicts of Interest: The authors declare no conflict of interest.

References

1. Gabor, D. A new microscopic principle. *Nature* **1948**, *161*, 777–778. [CrossRef]
2. Hendry, D. *Digital Holography: Digital Hologram Recording, Numerical Reconstruction and Related Techniques*; Schnars, U., Jueptner, W., Eds.; Springer: Berlin/Heidelberg, Germany, 2005; ISBN 354021934-X.
3. Lee, W.H. Sampled Fourier transform hologram generated by computer. *Appl. Opt.* **1970**, *9*, 639–643. [CrossRef]
4. Considine, P.S. Effects of coherence on imaging systems. *J. Opt. Soc. Am.* **1966**, *56*, 1001–1009. [CrossRef]
5. Mills, J.P.; Thompson, B.J. Effect of aberrations and apodization on the performance of coherent optical systems. II. Imaging. *J. Opt. Soc. Am.* **1986**, *3*, 704–716. [CrossRef]
6. Rosen, J.; Brooker, G. Digital spatially incoherent Fresnel holography. *Opt. Lett.* **2007**, *32*, 912–914. [CrossRef] [PubMed]
7. Vijayakumar, A.; Kashter, Y.; Kelner, R.; Rosen, J. Coded aperture correlation holography—A new type of incoherent digital holograms. *Opt. Express* **2016**, *24*, 12430–12441. [CrossRef] [PubMed]
8. Wu, J.; Zhang, H.; Zhang, W.; Jin, G.; Cao, L.; Barbastathis, G. Single-shot lensless imaging with fresnel zone aperture and incoherent illumination. *Light Sci. Appl.* **2020**, *9*, 53. [CrossRef] [PubMed]

9.   Naik, D.N.; Ezawa, T.; Singh, R.K.; Miyamoto, Y.; Takeda, M. Coherence holography by achromatic 3-D field correlation of generic thermal light with an imaging Sagnac shearing interferometer. *Opt. Express* **2012**, *20*, 19658–19669. [CrossRef] [PubMed]
10.   Naik, D.N.; Pedrini, G.; Osten, W. Recording of incoherent-object hologram as complex spatial coherence function using Sagnac radial shearing interferometer and a Pockels cell. *Opt. Express* **2013**, *21*, 3990–3995. [CrossRef] [PubMed]
11.   Roy, M.; Svahn, P.; Cherel, L.; Sheppard, C.J. Geometric phase-shifting for low-coherence interference microscopy. *Opt. Lasers Eng.* **2002**, *37*, 631–641. [CrossRef]
12.   Hariharan, P.; Oreb, B.F.; Eiju, T. Digital phase-shifting interferometry: A simple error-compensating phase calculation algorithm. *Appl. Opt.* **1987**, *26*, 2504–2506. [CrossRef] [PubMed]

Proceeding Paper

Roles of Deep Learning in Optical Imaging [†]

Vineela Chandra Dodda [1] and Inbarasan Muniraj [1,2,*]

[1] Department of Electronics and Communication Engineering, School of Engineering and Applied Sciences, SRM University AP, Amaravathi 522240, India; vineelachandra_dodda@srmap.edu.in

[2] LiFE Laboratory, Department of Electronics and Communication Engineering, Alliance College of Engineering and Design, Alliance University, Bengaluru 562106, India

* Correspondence: inbarasan.muniraj@alliance.edu.in

† Presented at the International Conference on "Holography Meets Advanced Manufacturing", Online, 20–22 February 2023.

Abstract: Imaging-based problem-solving approaches are an exemplary way of handling problems in various scientific applications. With an increased demand for automation, artificial intelligence techniques have shown exponential growth in recent years. In this context, deep-learning-based "learned" solutions have been widely adopted in many applications and are thus slowly becoming an inevitable alternative tool. It is known that in contrast to the conventional "physics-based" approach, deep learning models are a "data-driven" approach, where the outcomes are based on data analysis and interpretation. Thus, deep learning approaches have been applied in several (optical and computational) imaging-based scientific problems such as denoising, phase retrieval, hologram reconstruction, and histopathology, to name a few. In this work, we present two deep-learning networks for 3D image denoising and off-focus voxel removal.

Keywords: optical 3D imaging; unsupervised denoising; off-focus removal; integral imaging

Citation: Dodda, V.C.; Muniraj, I. Roles of Deep Learning in Optical Imaging. *Eng. Proc.* **2023**, *34*, 6. https://doi.org/10.3390/HMAM2-14123

Academic Editor: Vijayakumar Anand

Published: 6 March 2023

1. Introduction

Integral imaging (II) is one of the passive three-dimensional (3D) imaging techniques invented by Gabriel Lippmann in 1908 [1] and has received wide attention, as the appli cations of II span several research problems in optical engineering research areas [2–4]. For instance, these include biomedicine, security, autonomous vehicles, and remote sensing, to name a few [5].

Advanced machine learning (ML) and deep learning (DL) algorithms have been shown to produce superior results in computer-vision-based applications. Thereafter, such approaches have also been extended to solve several problems in various other scientific research areas. In particular, the DL framework has been proven as an important tool to make automatic decisions, as it solves numerous image-based problems without much human intervention. Convolution Neural Networks (CNN) are a widely used DL algorithm for several problems such as image classification [6], autonomous driving [7], etc. Furthermore, a CNN framework for 3D face recognition and classification in a photon-starved environment has also been demonstrated [2,8].

2. Integral Imaging

Integral imaging (II) captures a 3D scene in the form of two-dimensional (2D) elemental images (EIs) in addition to the directional information (i.e., angle of propagation). Notably, 3D scene reconstruction can be achieved in two ways: (i) optical methods and (ii) computational methods [9]. In computational integral imaging (CII), a geometric ray back-propagation method is employed which magnifies and superimposes the EIs onto each other to reconstruct 3D sectional images [10]. Consequently, the objects or 3D points which are located at the corresponding depth position in an imaging plane are properly overlapped and in focus, while the other points at different depth locations do not overlap properly and hence appear off-focus or defocused. The defocused points in the 3D sectional image do not convey any valuable information and are therefore redundant. Recently, we have demonstrated a way to manually identify and remove the off-focus points from a 3D sectional image [11]. Furthermore, under some special imaging scenarios (e.g., biomedical imaging and night vision), low light levels or photon-starved illumination conditions may be encountered. In such cases, since image capturing happens in much darker conditions, the recorded image looks degraded due to the presence of noise [8,10]. Nevertheless, this system has been shown to provide a better 3D reconstruction in terms of the PSNR even with fewer photons, e.g., 100 photons [10].

2.1. Denoising

For image denoising, various methods have been proposed in the literature such as prediction filtering, transformation-based methods, rank reduction methods, and dictionary learning methods, to name a few. In addition to these, DL algorithms have also been applied to the image denoising problem [12]. In this regard, there are two methods that are commonly followed to train the DL network: (i) supervised and (ii) unsupervised. First, we discuss supervised learning, where an under-complete autoencoder is used to denoise the noisy 3D integral (sectional) images with a patch-based approach. In this process, the noisy input 3D sectional image is divided into multiple patches, which are then used to train the neural network in a supervised manner (we use clean data as labels). We note that by using the patch-based approach, the time required to prepare the labeled training data is greatly reduced. Then, after denoising, the acquired denoised patches can be combined via an unpatching process. Figure 1 depicts the supervised denoising technique used on our dataset [13]. To train the network, 20 epochs were employed with a learning rate of 0.001.

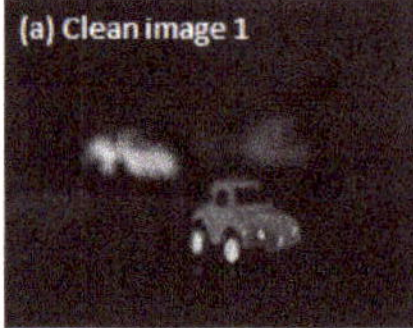

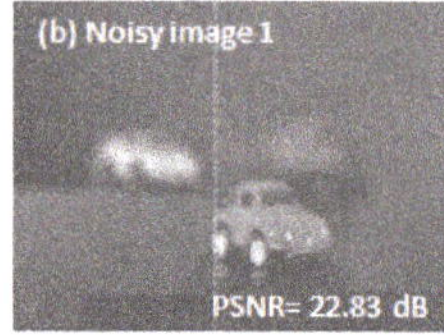

Figure 1. Denoised results for supervised learning.

Figure 1c shows the denoised 3D sectional image. We analyzed the performance of the proposed method quantitatively in terms of the peak signal-to-noise ratio (PSNR). For instance, the PSNR value given in Figure 1c is an estimation from Figure 1a,c. It is evident from Figure 1c that the proposed denoising method has a better performance in terms of the PSNR. Second, we proposed an unsupervised learning method for 3D image denoising. In this study, we opted for a U-Net architecture [8]. This is an end-to-end, fully unsupervised denoising approach where the noisy photons in the 3D sectional image are fed as an input to the network. The major components in the U-Net are encoder and decoder blocks with skip connection layers [14–16]. In addition to this, skip blocks (SB) were added to the skip connection strategy in the U-Net architecture to avoid the vanishing gradients problem. In the training process, the 3D input image is given in the form of patches to the network. The patched input image is converted to a 1D vector and fed as an input to the network. After removing the noise, we unpatch the 1D vector and convert it back to the size of the input data. In our experiments, to test the performance of the proposed method, we used two 3D objects: a tri-colored ball known as Object 1 in Figure 2a and a toy bird referred to as Object 2 in Figure 2(a1,a2). Figure 2(b1,b2) are obtained after the TV denoising method. The proposed method results are given in Figure 2(c1,c2). Notably, we used 20% of the PCSI patches for validation and 60% of the patches for training purposes. In this work, 15 epochs were used with a learning rate of 0.001 to train the network. The PSNR values are shown in Figure 2.

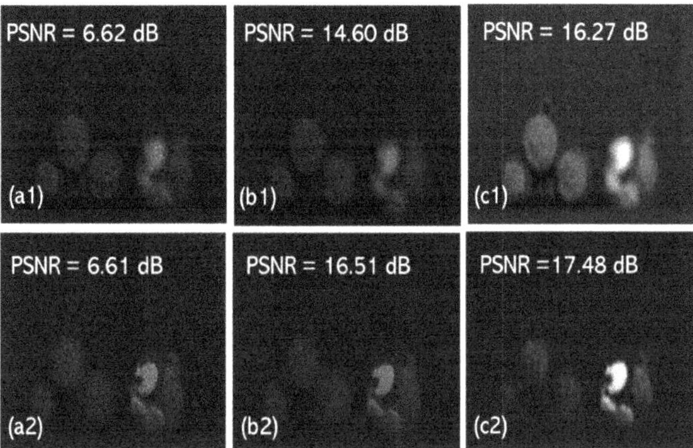

Figure 2. Denoising results: (**a1,b1,c1**) represent the noisy photon-counted 3D sectional image, the TV denoised image, and the result of our proposed denoising method when object 1 is in focus, respectively, and (**a2,b2,c2**) represent the noisy photon-counted 3D sectional image, the TV denoised image, and the result of our proposed denoising method when object 2 is in focus, respectively.

2.2. Off-Focus Removal

Several studies have been conducted to demonstrate the feasibility of combining photon detection imaging or photon counting imaging (PCI) techniques with conventional 3D integral imaging systems, known as photon counted integral imaging (PCII) [2,9–11,17]. In such systems, it is known that the reconstructed depth images contain both the focused and off-focus (or out-of-focus) voxels simultaneously (see for instance Figure 3). Off-focus pixels often look blurred and therefore do not convey clear information about the scene. Several approaches have been proposed to efficiently remove the off-focus points from the reconstructed 3D images [4,11]. We note that the existing approaches are subjective as they involve manual calculation of algorithm parameters such as variance, threshold, etc., which is time consuming.

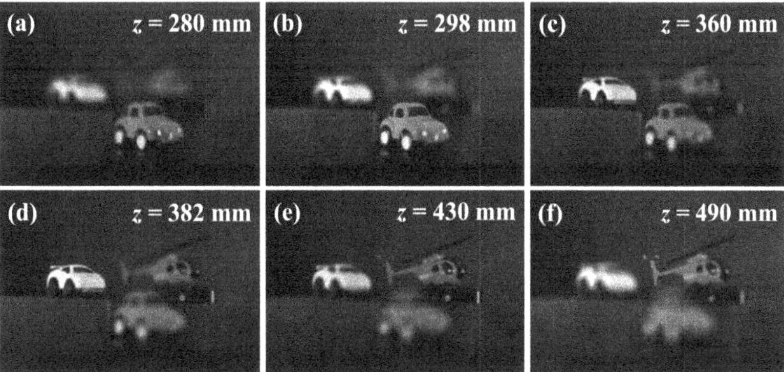

Figure 3. Reconstructed 3D CII sectional images at various depth locations.

Here, we propose a new ensemble Dense Neural Network (DNN) model that is composed of six different DNN models, each trained with its own set of training datasets for removing off-focus points from 3D sectional images. It is known that data pre-processing enhances the accuracy of the network; therefore, we used the Otsu thresholding algorithm [18] to remove the unwanted (and obvious) background from the 3D sectional images. In this work, we employed an ADAM optimizer to update the weights and bias [19], and the standard mean squared error (MSE) was used as the cost function in our training process. Notably, the proposed ensemble deep neural network was trained (supervised way) using the conventional 3D sectional images from various depth locations and the corresponding focused images (labels). We tested the method on a 3D scene that contains two toy cars and one toy helicopter (see Figure 4) [13]. We used an Intel® Xeon® Silver 4216 CPU @2.10 GHz (two processors) with 256 GB RAM and a 64-bit operating system to simulate all the scenarios.

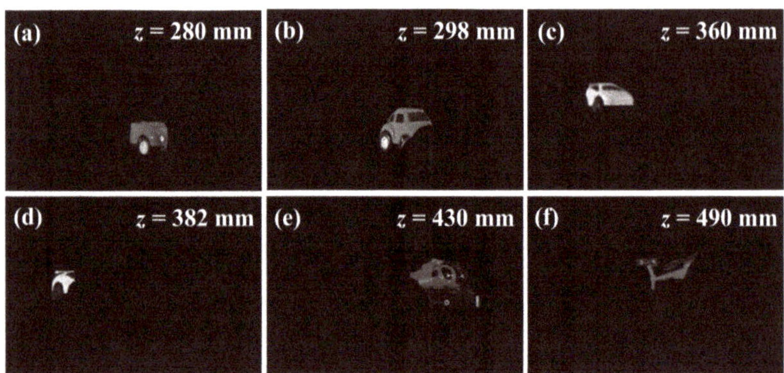

Figure 4. Reconstructed focus only CII sectional images by using the proposed DL network.

3. Conclusions

In summary, we demonstrated that it is possible to use deep learning networks to solve some of the inherent problems of 3D optical imaging systems. For instance, we have tackled two important problems that exist in 3D integral imaging systems, i.e., denoising and off-focus removal, using two different datasets. For our study, it is evident that DL can be used to solve problems that are too complex to carry out manually. It is therefore expected that we will further expand our analysis to various other imaging modalities such as holography, microscopy, etc.

Author Contributions: V.C.D. and I.M. contributed equally to manuscript preparation. All authors have read and agreed to the published version of the manuscript.

Funding: This research was funded by the Department of Science and Technology (DST) under the Science and Engineering Research Board (SERB) grant number SRG/2021/001464.

Institutional Review Board Statement: Not applicable.

Informed Consent Statement: Not applicable.

Data Availability Statement: Data for this paper is not publicly available but shall be provided upon reasonable request to the corresponding author.

Acknowledgments: Authors thank Suchit Patel of Poornima College of Engineering, India, for lending his support in the simulations and we sincerely thank Bahram Javidi of the University of Connecticut and Moon Inkyu of DGIST, Korea, for providing the dataset.

Conflicts of Interest: The authors declare no conflict of interest.

References

1. Lippmann, G. La photographie integrale. *Comptes-Rendus* **1908**, *146*, 446–451.
2. Markman, A.; Javidi, B. Learning in the dark: 3D integral imaging object recognition in very low illumination conditions using convolutional neural networks. *OSA Contin.* **2018**, *1*, 373–383. [CrossRef]
3. Yeom, S.; Javidi, B.; Watson, E. Three-dimensional distortion-tolerant object recognition using photon-counting integral imaging. *Opt. Express* **2007**, *15*, 1513–1533. [CrossRef] [PubMed]

4. Yi, F.; Lee, J.; Moon, I. Simultaneous reconstruction of multiple depth images without off-focus points in integral imaging using a graphics processing unit. *Appl. Opt.* **2014**, *53*, 2777–2786. [CrossRef] [PubMed]
5. Javidi, B.; Carnicer, A.; Arai, J.; Fujii, T.; Hua, H.; Liao, H.; Martínez-Corral, M.; Pla, F.; Stern, A.; Waller, L.; et al. Roadmap on 3D integral imaging: Sensing, processing, and display. *Opt. Express* **2020**, *28*, 32266–32293. [CrossRef] [PubMed]
6. Zhang, K.; Zhang, Z.; Li, Z.; Qiao, Y. Joint face detection and alignment using multitask cascaded convolutional networks. *IEEE Signal Process. Lett.* **2016**, *23*, 1499–1503. [CrossRef]
7. Chen, C.; Seff, A.; Kornhauser, A.; Xiao, J. Deepdriving: Learning affordance for direct perception in autonomous driving. In Proceedings of the IEEE International Conference on Computer Vision, Santiago, Chile, 7–13 December 2015; pp. 2722–2730.
8. Dodda, V.C.; Kuruguntla, L.; Elumalai, K.; Chinnadurai, S.; Sheridan, J.T.; Muniraj, I. A denoising framework for 3D and 2D imaging techniques based on photon detection statistics. *Sci. Rep.* **2023**, *13*, 1365. [CrossRef] [PubMed]
9. Moon, I.; Muniraj, I.; Javidi, B. 3D visualization at low light levels using multispectral photon counting integral imaging. *J. Disp. Technol.* **2013**, *9*, 51–55. [CrossRef]
10. Muniraj, I.; Guo, C.; Lee, B.G.; Sheridan, J.T. Interferometry based multispectral photon-limited 2D and 3D integral image encryption employing the Hartley transform. *Opt. Express* **2015**, *23*, 15907–15920. [CrossRef] [PubMed]
11. Muniraj, I.; Guo, C.; Malallah, R.; Maraka, H.V.R.; Ryle, J.P.; Sheridan, J.T. Subpixel based defocused points removal in photon-limited volumetric dataset. *Opt. Commun.* **2017**, *387*, 196–201. [CrossRef]
12. Choi, G.; Ryu, D.; Jo, Y.; Kim, Y.S.; Park, W.; Min, H.S.; Park, Y. Cycle-consistent deep learning approach to coherent noise reduction in optical diffraction tomography. *Opt. Express* **2019**, *27*, 4927–4943. [CrossRef] [PubMed]
13. Dodda, V.C.; Kuruguntla, L.; Elumalai, K.; Muniraj, I.; Chinnadurai, S. An undercomplete autoencoder for denoising computational 3D sectional images. In *Computational Optical Sensing and Imaging*; Optica Publishing Group: Washington, DC, USA, 2022; p. JW2A-19.
14. Lempitsky, V.; Vedaldi, A.; Ulyanov, D. Deep image prior. In Proceedings of the 2018 IEEE/CVF Conference on Computer Vision and Pattern Recognition, Salt Lake City, UT, USA, 18–23 June 2018; pp. 9446–9454.
15. Yang, L.; Wang, S.; Chen, X.; Saad, O.M.; Chen, W.; Oboué, Y.A.S.I.; Chen, Y. Unsupervised 3-D Random Noise Attenuation Using Deep Skip Autoencoder. *IEEE Trans. Geosci. Remote Sens.* **2021**, *60*, 5905416. [CrossRef]
16. Kingma, D.P.; Ba, J. Adam: A method for stochastic optimization. *arXiv* **2014**, arXiv:1412.6980.
17. Tavakoli, B.; Javidi, B.; Watson, E. Three dimensional visualization by photon counting computational integral imaging. *Opt. Express* **2008**, *16*, 4426–4436. [CrossRef] [PubMed]
18. Otsu, N. A threshold selection method from gray-level histograms. *IEEE Trans. Syst. Man Cybern.* **1979**, *9*, 62–66. [CrossRef]

Digital Fourier Transform Holography Using a Beam Displacer [†]

Mohit Rathor *, Shivam Kumar Chaubey and Rakesh Kumar Singh

Laboratory of Information Photonics and Optical Metrology, Department of Physics, Indian Institute of Technology (Banaras Hindu University), Varanasi 221005, Uttar Pradesh, India; shivamkumarchaubey.rs.phy21@itbhu.ac.in (S.K.C.); krakeshsingh.phy@iitbhu.ac.in (R.K.S.)
* Correspondence: mohitrathor.rs.phy21@itbhu.ac.in
† Presented at the International Conference on "Holography Meets Advanced Manufacturing", Online, 20–22 February 2023.

Abstract: Fourier transform holography overcomes the phase recovery challenge through recording complex field information of the object in an interference pattern recorded at the far field, i.e., Fourier plane. Moreover, this geometry helps to reconstruct the complex field of the object from a single Fourier transform, which is an attractive feature for the numerical reconstruction of the digitally recorded hologram. In this paper, we present a nearly common path experimental design for recording a digital Fourier holographic hologram using a beam displacer, and recover the complex valued objects using the Fourier analysis. The performance of the system is experimentally examined for different objects.

Keywords: digital Fourier transform holography; phase imaging; common path configuration

Citation: Rathor, M.; Chaubey, S.K.; Singh, R.K. Digital Fourier Transform Holography Using a Beam Displacer. *Eng. Proc.* **2023**, *34*, 7. https://doi.org/10.3390/HMAM2-14126

Academic Editor: Vijayakumar Anand

Published: 6 March 2023

1. Introduction

Phase information of the objects in imaging systems are usually extracted to estimate the optical features of transparent objects, such as cells, glasses, optical elements and any transparency objects. The transparent structure and topology of the object can be estimated using the refractive index difference between the object measured and its neighboring media. Thus, quantitative phase information of the object plays a vital role to explain the realistic features of the object. However, phase is not a directly observable quantity in the optical domain due to the high frequency of the wave.

Where retrieving the phase information of an object is concerned, digital holography (DH) is a useful emerging techniques due to its capability to record and reconstruct complex fields [1,2]. DH appears as a promising computational and quantitative 3D imaging technique, being rooted in interferometry of the coherent light and numerical reconstruction of the optically recorded hologram [3]. Quantitative information of the object is recorded as an interference fringe pattern using a digital camera. Different experimental techniques were previously proposed for the DH and major designs are the on-axis, off-axis and phase shifting. The on-axis, or Gabor's, geometry is compact and stable; however, it faces the challenge of the twin image issue and overlap of the spectra [4]. Off-axis holography overcomes these limitations at the cost of angular separation between the interfering beams [5,6]. Different experimental designs were previously developed to provide the angular separation between the interfering beams at the cost of strict requirement of vibration isolation in the experimental design [7]. Nevertheless, an-off axis holography scheme is highly desired due to its capability to recover the quantitative image from a single measurement in contrast to phase shifting and iteration-free nature. Moreover, the cost of numerical reconstruction of the hologram can be reduced using the Fourier transform geometry in the off-axis holography [8].

In this paper, we demonstrate a new experimental design for recording the Fourier transform holography (FTH). This outcome is achieved through making an experimental design using the beam displaces to spatially separate the orthogonal polarization components of the light source. Subsequently, we used an appropriate method to interfere with these polarization components, with one loaded with the object and other used as a reference, to record an off-axis hologram. The phase information of the object is encoded into the interference pattern recorded at the far field. Our experimental setup uses nearly common paths for the interfering waves; hence, the experimental design is stable. The performance of the system is experimentally examined for different objects; the results are shown below.

2. Experimental Setup

A schematic design of the proposed system is shown in Figure 1. A horizontal polarized coherent light source from a He-Ne laser with an optical wavelength of 632.8 nm (model no.: HNL150LB; makes: Thorlabs) is used and collimated using a spatial filter assembly composed of the microscope objective (MO), pinhole, and a bi-convex lens L1. This collimated light beam is turned to a diagonal polarization using a half-wave plate (HWP) placed at an angle of 22.5° from the horizontal pass axis. The diagonally polarized beam splits into two orthogonal polarized components (horizontal and vertical) after passing through a beam displacer (model no.: BD40; makes: Thorlabs). One of the polarization components, for example horizontal, passes through an object placed at the front focal plane of lens L2, as shown in Figure 1. This beam works as an object beam, which is transversely separated from the vertically polarized component. On the other hand, a parallel propagating vertically polarized beam is filtered using a pinhole placed at the same transverse plane as the object plane at a distance f from the L2. This beam works as the reference beam. Lens L2 performs the Fourier transform of both these beams, and the interference pattern is recorded through placing a polarizer at 45° from the pass axis of the horizontal in front of the CMOS camera (model no.: DCC3240M; makes: Thorlabs). A complete information of the complex-valued object (i.e., amplitude and phase information) is encoded into the recorded Fourier transform of the hologram. In order to reconstruct the recorded hologram, we used a single fast Fourier transform. The complex information of the objects is shown below.

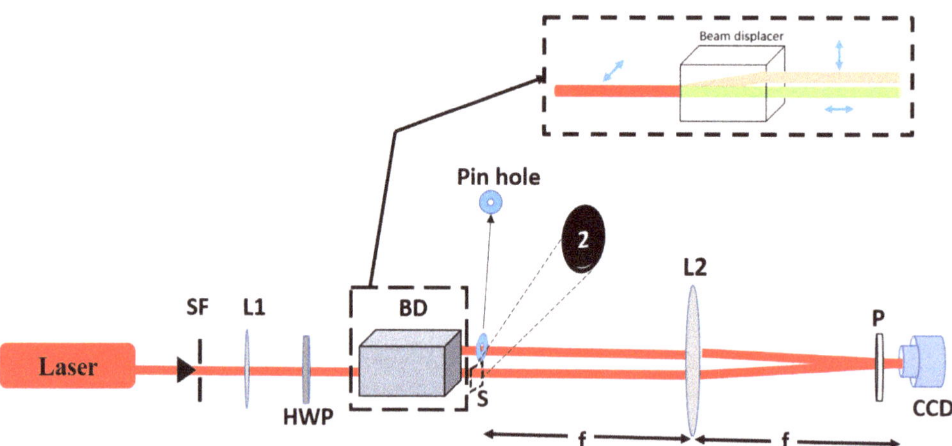

Figure 1. A compact experimental setup design for recording digital Fourier transform hologram with help of a beam displacer. HWP is a half wave-plate, SF is a spatial filter, L1 and L2 are lenses, BD is beam displacer, S is sample, P is a polarizer and CMOS is complementary metal oxide semiconductor camera.

3. Result and Discussion

To demonstrate the appropriateness of our technique, we used two different transparency objects (2 and ψ). Figure 2 shows recorded holograms corresponding to these objects.

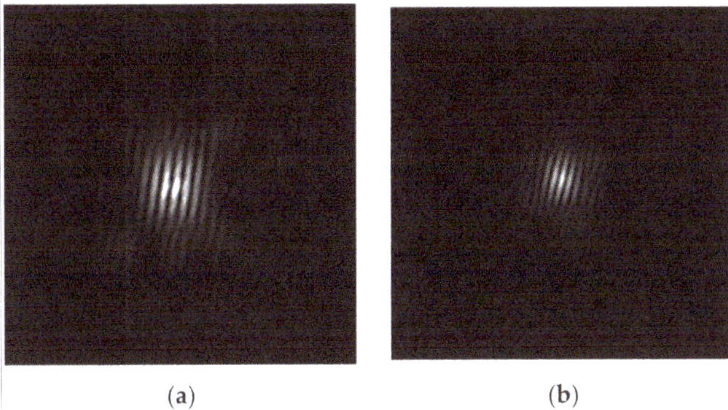

Figure 2. Fourier transform hologram of object (**a**) 2; (**b**) ψ.

Reconstruction of the complex valued objects from these two holograms are shown in Figure 3.

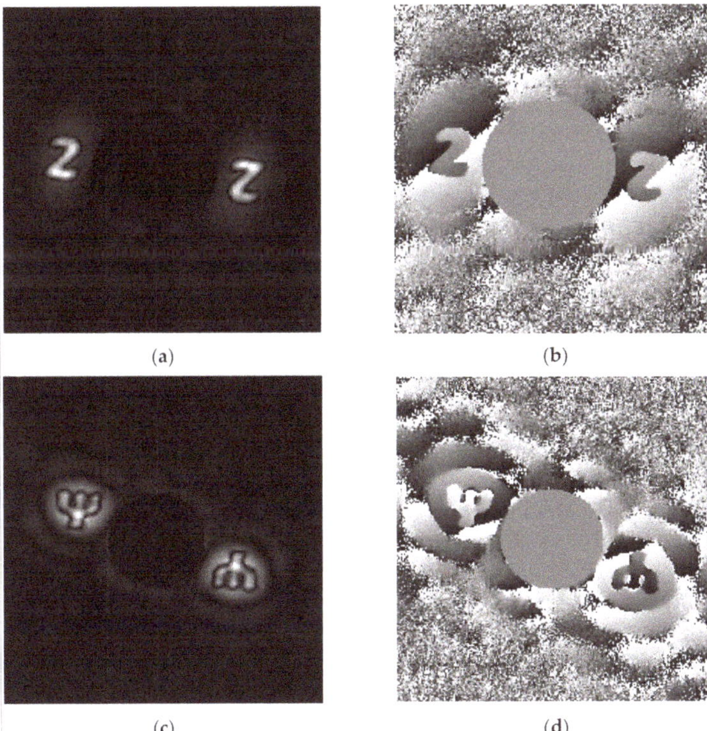

Figure 3. (**a**,**b**) show amplitude and phase information of object 2, and (**c**,**d**) show amplitude and phase information of object infrromation ψ.

A Fourier transform of the hologram generates three dominant frequency regions, as shown in Figure 3. The central part, i.e., the unmodulated components, is strong and suppressed in order to highlight the reconstructed object and its conjugate in the off-axis position. The off-axis location of the reconstructed object in the frequency domain is decided using the transverse spatial separation between the interfering beams, as shown in Figure 3a,b for object 2. Similarly, the reconstruction of object ψ is shown in Figure 3c,d.

The experimental result of the objects (2 and ψ) validates the performance of the proposed design of the holographic setup. The proposed design of this experimental setup provides a nearly common path configuration system using a beam displacer, which makes the system more stable in comparison to experimental designs based on the Michelson and Mach–Zehnder configurations.

4. Conclusions

In this paper, we present a nearly common path experimental configuration system using a beam displacer for recording a digital Fourier transform hologram, while the complex-valued objects information are presented using the fast Fourier analysis. Moreover, this geometry helps to reconstruct the complex field of the object from a single Fourier transform. The performance of the system is experimentally examined for different objects.

Author Contributions: Idea conception, manuscript preparation and methodology design, M.R.; manuscript preparation, methodology design and experimental design, S.K.C.; advice, review, supervision and editing, R.K.S. All authors were involved in supervision, formulation of research goals, funding acquisition, reviewing and editing. All authors have read and agreed to the published version of the manuscript.

Funding: This research was funded by Department of Biotechnology (DBT) grant no. BT/PR/35587/MED/32/707/2019 and Council of Scientific and Industrial Research, India (CSIR India- Grant No.: 80 (0092)/20/EMR-II).

Institutional Review Board Statement: Not applicable.

Informed Consent Statement: Not applicable.

Data Availability Statement: The data used in this paper are not currently publicly accessible, although they are available from the authors upon reasonable request.

Acknowledgments: Mohit Rathor would like to acknowledge IIT (BHU), Varanasi.

Conflicts of Interest: The authors declare no conflict of interest.

References

1. Grilli, S.; Ferraro, P.; De Nicola, S.; Finizio, A.; Pierattini, G.; Meucci, R. Whole optical wavefields reconstruction by digital holography. *Opt. Express* **2001**, *9*, 294–302. [CrossRef] [PubMed]
2. Goodman, J.W. *Introduction of Fourier Optics*, 2nd ed.; McGraw-Hill: New York, NY, USA, 1996; pp. 295–363.
3. Samsheerali, P.T.; Khare, K.; Joseph, J. Quantitative phase imaging with single shot digital holography. *Opt. Commun.* **2014**, *319*, 85–89. [CrossRef]
4. Lecture, N. Gabor-Lecture-Holography. *Science* **1972**, *177*, 299.
5. Liebling, M.; Blu, T.; Unser, M. Complex-wave retrieval from a single off-axis hologram. *JOSA A* **2004**, *21*, 367–377. [CrossRef] [PubMed]
6. Cuche, E.; Marquet, P.; Depeursinge, C. Spatial filtering for zero-order and twin-image elimination in digital off-axis holography. *Appl. Opt.* **2000**, *39*, 4070–4075. [CrossRef] [PubMed]
7. Shaked, N.T. Quantitative phase microscopy of biological samples using a portable interferometer. *Opt. Lett.* **2012**, *37*, 2016–2018. [CrossRef] [PubMed]
8. Apostol, D.; Sima, A.; Logofătu, P.C.; Garoi, F.; Damian, V.; Nascov, V.; Iordache, I. Fourier Transform Digital Holography. In Proceedings of the OMOPTO 2006: Eighth Conference on Optics, Sibiu, Romania, 1 August 2007. [CrossRef]

Proceeding Paper

An Asymmetric Optical Cryptosystem Using Physically Unclonable Functions in the Fresnel Domain [†]

Vinny Cris Mandapati [1,*], Shashi Prabhakar [2], Harsh Vardhan [1], Ravi Kumar [1], Salla Gangi Reddy [1], Sakshi [3] and Ravindra P. Singh [2]

[1] Department of Physics, SRM University—AP, Amaravati 522502, India; harshvardhan_r@srmap.edu.in (H.V.); ravi.k@srmap.edu.in (R.K.); gangireddy.s@srmap.edu.in (S.G.R.)
[2] Physical Research Laboratory, Navrang Pura, Ahmedabad 380009, India; shaship@prl.res.in (S.P.); rpsingh@prl.res.in (R.P.S.)
[3] Department of Chemical Engineering, Ben-Gurion University of the Negev, P.O. Box 653, Beer-Sheva 8410501, Israel; sakshich289@gmail.com
* Correspondence: vinnycris_m@srmap.edu.in
† Presented at the International Conference on "Holography Meets Advanced Manufacturing", Online, 20–22 February 2023.

Abstract: In this paper, we propose a new asymmetric cryptosystem for phase image encryption, using the physically unclonable functions (PUFs) as security keys. For encryption, the original amplitude image is first converted into a phase image and modulated with a PUF to obtain a complex image. This complex image is then illuminated with a plane wave, and the complex wavefront at a distance d is recorded. The real part of the complex wavefront is further processed to obtain the encrypted image and the imaginary part is kept as the private key. The polar decomposition approach is utilized to generate two more private security keys and to enable the multi-user capability in the cryptosystem. Numerical simulations confirm the feasibility of the proposed method.

Keywords: asymmetric cryptosystems; physically unclonable functions; polar decomposition method

Citation: Cris Mandapati. V.; Prabhakar, S.; Vardhan, H.; Kumar, R.; Reddy, S.G.; Sakshi; Singh, R.P. An Asymmetric Optical Cryptosystem Using Physically Unclonable Functions in the Fresnel Domain. *Eng. Proc.* **2023**, *34*, 8. https://doi.org/10.3390/HMAM2-14124

Academic Editor: Vijayakumar Anand

Published: 6 March 2023

1. Introduction

The advancements in data transfer and storage technology have prompted new challenges in guaranteeing its secure transmission. A huge amount of data (images, passwords, bank details, etc.) is transmitted daily through open channels, making it vulnerable to intruders. To ensure safe transmission, several optical and digital encryption techniques have been explored. The first optical cryptosystem, i.e., the double-random phase-encoding (DRPE) technique, was demonstrated back in 1995. This employs a simple 4-f setup to encode a two-dimensional image into a white noise-like distribution [1]. With time, many variants of DRPE in other transform domains, such as fractional Fourier, gyrator, Mellin, Hartley, wavelet, and cosine transforms, have been explored to improve data transmission security [2,3]. Various other optical approaches like diffractive imaging and interference methods, as well as polarization encoding, were also explored to design new and sophisticated optical cryptosystems [2]. Most of these methods are symmetric in nature and vulnerable to different kinds of attacks [3,4]. To resist these attacks, several attempts were made to develop asymmetric cryptosystems that offer nonlinearity and serve as secure options against well-known attacks, such as known plaintext attacks, chosen plaintext attacks and cipher text-only attacks [3]. Although several cryptosystems have been developed in recent years, the search continues for newer advanced methods which can provide better security with less practical complexity and are computationally efficient.

In this paper, we present a new asymmetric cryptosystem using physically unclonable functions as security keys. Mostly, the security keys used in the existing image encryption methods are computer-generated noise-like distributions with uniformly distributed histograms. On the other hand, the PUFs used in this paper are relatively unbreakable non-algorithmic functions which are difficult to reproduce. PUFs carry unique signatures due to the random and stochastic processes involved in their generation [5]. The PUFs employed here are obtained as random speckles from the coherence light source when passed through a ground glass diffuser. Statistically, an attacker may retrieve the computer-generated keys if they have partial/full knowledge of the cryptosystem; however, since PUFs are generated physically, it is difficult to retrieve them through iterative algorithms.

2. Proposed Technique and Results

The proposed technique is designed in the Fresnel transform domain to make it lensless. Polar decomposition (PD) process aids in making the system asymmetric and enables multiuser capabilities [4]. PUFs are generated experimentally, as discussed in Ref. [5]. PD essentially factorizes the input matrix into a set of linearly independent matrices, i.e., one rotational matrix and two symmetric matrices. To reconstruct the input matrix or image, only the rotational matrix and one of the symmetric matrices is required [4].

For encryption, the first input image, $f(x, y)$, is phase-encoded as $\exp(i\pi f(x, y))$ and modulated with the first PUF phase function as $A(x, y) = \exp(i\pi f(x, y)) \times PUF1$. The complex image $A(x, y)$ is then Fresnel-propagated with distance d_1, to obtain the complex wavefront $A'(u, v)$. Next, the real and imaginary parts of $A'(u, v)$ are separated; the real part undergoes the polar decomposition process; and the imaginary part is retained as the first private key. The PD will result in three images, i.e., R: the rotational image; and U and V: the positive symmetric matrix images. U and V are stored as the private keys and can be distributed to two different users for individual decryption. The rotational matrix part $R(u, v)$ is further Fresnel-propagated to a distance d_2, which results in the complex wavefront $B(u', v')$. This complex amplitude image is then modulated with the second PUF2 to obtain the final encrypted image $E(u', v')$. The original image can be recovered through the reverse process using all the correct keys. The flowchart of the encryption and decryption process is illustrated in Figure 1a,b, respectively.

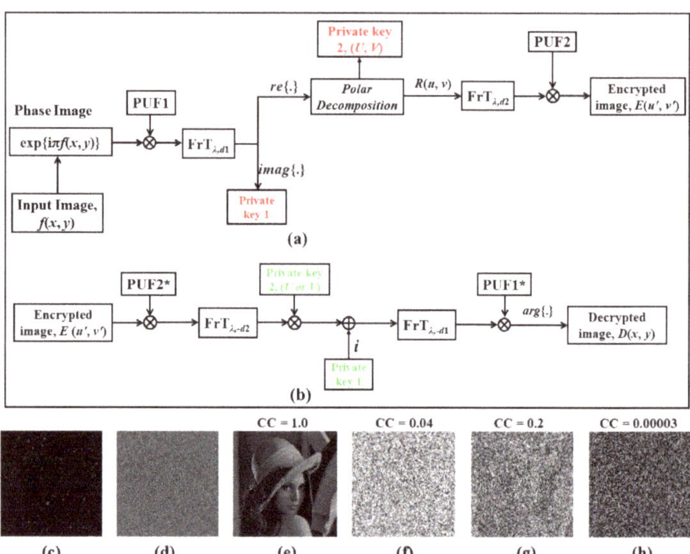

Figure 1. Flowchart for (**a**) encryption process; (**b**) decryption process; (c) PUF used (order of optical vortex = 2); (**d**) encrypted image. Decrypted image with (**e**) all correct keys (**f**) deviation in d_1 = 2 mm (**g**) deviation in d_2 = 2 mm (**h**) using wrong PUF (order of optical vortex = 3). The red colored details denotes private keys 1 and 2 in encryption process and green colored details denotes the keys in decryption process; '*' represents the complex conjugate of the function.

The validity of the proposed technique was verified by performing numerical simulations on MATLAB^{TM} (Mathworks, Natick, MA, USA, version 2022(b)) on an AMD Ryzen 5-5500 U laptop, with 16 GB RAM. The 'Lena' image of 256 × 256 pixels is used as the input image. Figure 1c shows one of the PUFs used for encryption and the final encrypted image is shown in Figure 1d, whereas the decrypted image with all correct keys is shown in Figure 1e. The sensitivity of security keys is also checked by performing decryption with a small deviation in Fresnel parameters or by using wrong keys. The corresponding results are shown in Figure 1f–h. The results confirm that the proposed method is feasible and sensitive to the keys.

3. Concluding Remarks

In conclusion, a new asymmetric optical cryptosystem with multiuser capabilities is proposed using polar decomposition in the Fresnel domain. The method has a large set of keys which include the Fresnel propagation parameters, two variable PUFs, and three private keys generated during the encryption process. The PUFs used as security keys are difficult to replicate, a fact which improves the robustness against various attacks. The sensitivity of all the keys is also verified. The work is a subject of our ongoing research and will be presented in detail in the near future.

Author Contributions: Conceptualization, S.P., R.K., S.G.R. and R.P.S.; methodology, V.C.M., H.V., S.P., S.G.R., S. and R.K.; software, V.C.M., H.V., S. and R.K.; validation, V.C.M. and H.V.; formal analysis, V.C.M. and S.; investigation, V.C.M., H.V. and S.; resources, S.G.R., S.P. and R.P.S.; data curation, V.C.M. and H.V.; writing—original draft preparation, V.C.M., H.V., R.K. and S.G.R.; writing—review and editing, R.K., S.G.R. and R.P.S.; visualization, V.C.M. and H.V.; supervision, R.K. and S.G.R.; project administration, R.K. and S.G.R. All authors have read and agreed to the published version of the manuscript.

Funding: The research was granted by the Science and Engineering Research Board (SERB), the Government of India, under the start-up research grant (Grant No. SERB/SRG/2019/000857), and SRM University—AP for seed research grants under SRMAP/URG/CG/2022-23/006 and SRMAP/URG/E&PP/2022-23/003 to S.G.R.

Institutional Review Board Statement: Not applicable.

Informed Consent Statement: Not applicable.

Data Availability Statement: The data related to the paper are available from the corresponding authors upon reasonable request.

Conflicts of Interest: The authors declare no conflict of interest.

References

1. Refregier, P.; Javidi, B. Optical Image Encryption Based on Input Plane and Fourier Plane Random Encoding. *Opt. Lett.* **1995**, *20*, 767–769. [CrossRef] [PubMed]
2. Nischal, K.N. *Optical Cryptosystems*; IOP Publishing Ltd.: Bristol, UK, 2020; pp. 2-1–2-18.
3. Javidi, B.; Carnicer, A.; Yamaguchi, M.; Nomura, T.; Pérez-Cabré, E.; Millán, M.S.; Markman, A. Roadmap on Optical Security. *J. Opt.* **2016**, *18*, 083001. [CrossRef]
4. Kumar, R.; Quan, C. Asymmetric multi-user optical cryptosystem based on polar decomposition and Shearlet transform. *Opt. Lasers Eng.* **2022**, *120*, 118–126. [CrossRef]
5. Vantiha, P.; Manupati, B.; Muniraj, I.; Anamalamudi, S.; Reddy, S.G.; Singh, R.P. Augmenting Data Security: Physical Unclonable Functions for linear canonical transform based cryptography. *Appl. Phys B* **2022**, *128*, 183. [CrossRef]

Proceeding Paper

Investigation of Memristor-Based Neural Networks on Pattern Recognition [†]

Gayatri Routhu [1], Ngangbam Phalguni Singh [1,*], Selvakumar Raja [2] and Eppala Shashi Kumar Reddy [1]

[1] Department of Electronics and Communication Engineering, Koneru Lakshmaiah Education Foundation, KL Deemed to be University, Vijayawada 522302, India; routhugayatri0207@gmail.com (G.R.); shashieppala150@gmail.com (E.S.K.R.)

[2] Department of Electronics and Communication Engineering, Anna University, Chennai 600025, India; selvakumariitb@live.com

* Correspondence: phalsingh@gmail.com

† Presented at the International Conference on "Holography Meets Advanced Manufacturing", Online, 20–22 February 2023.

Abstract: Mobile phones, laptops, computers, digital watches, and digital calculators are some of the most used products in our daily life. In the background, to make these gadgets work as per our desire, there are many simple components necessary for electronics to function, such as resistors, capacitors, and inductors, which are three basic circuit elements. The Memristor is one such component. This paper provides simulation results of the memristor circuit and its V-I characteristics at different functions as an input signal. A well-trained ANN is able to recognize images with higher precision. To enhance the properties such as accuracy, precision, and efficiency in recognition, memristor characteristics are introduced to the neural network, however, older devices experience some non-linearity issues, causing conductance-tuning problems. At the same time, to be used in some advanceable applications, ANN requires a huge amount of vector-matrix multiplication based on in-depth network expansion. An ionic floating gate (IFG) device with the characteristics of a memristive device can solve these problems. This work proposes a fully connected ANN using the IFG model, and the simulation results of the IFG model are given as synapses in deep learning. We use algorithms such as the gradient-descent model, forward and backward propagation for network building, and weight setting in neural networks to enhance their ability to recognize images. A well-trained network is formed by tuning those memristive devices to an optimized state. The synaptic memory obtained from the IFG device will be used in other deep neural networks to increase recognition accuracy. To be an activation function in the neural network, sigmoid functions were used but later replaced by the ReLu function to avoid vanishing gradients. This paper shows how images were recognized by their front, top, and side views.

Keywords: memristor; memristive device; ionic floating gate (IFG); artificial neural network (ANN); gradient descent model; forward and backward propagation; ReLu function

Citation: Routhu, G.; Phalguni Singh, N.; Raja, S.; Reddy, E.S.K. Investigation of Memristor-Based Neural Networks on Pattern Recognition. *Eng. Proc.* **2023**, *34*, 9. https://doi.org/10.3390/HMAM2-14149

Academic Editor: Vijayakumar Anand

Published: 13 March 2023

1. Introduction

A memristor is a simple passive two-terminal structure first coined by Professor Leon Chua in 1971. It is the concatenation of a "Memory resistor", which is a fourth fundamental circuit element besides resistor, capacitor, and inductor; this is the missing pair link that states the relationship between charge and flux over four basic circuit variables [1]. Depending on the input signal applied, there are two types of memristors: current-controlled and voltage-controlled [2]. The basic memristor is given in Figure 1.

Using Equations (1) and (2), we can find the value of memristance or memductance. The properties of the memristor made it useful for non-volatile memory and storage technology applications as it can theoretically develop multiple states. It can also operate at a very low voltage level. For the operation of a neural network as a synapse, these properties of memristor could be useful. Equations (1) and (2) below show the relationship between the flux and charge concerning voltage and current [1].

$$M(q) = \frac{d\Psi}{dq} \tag{1}$$

$$M(q(t)) = \frac{d\Psi}{dt} \cdot \frac{dt}{dq} = \frac{V(t)}{I(t)} \tag{2}$$

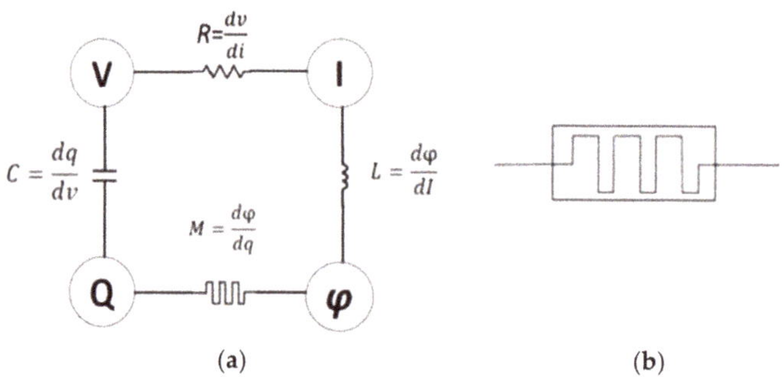

(a)   (b)

Figure 1. (**a**) The link between the voltage and current is shown by a resistor, voltage, and charge by a capacitor, and current and flux by an inductor; (**b**) Memristor basic symbol.

ANNs are interconnected groups of nodes that are composed of artificial neurons. An artificial neural network is a non-linear and self-adapting computational model that retains the biological concept of neural networks as human/animal brains [3]. An ANN has three layers, an input layer, a hidden layer, and the output layer. Every node of the input layer connects through links with every node of the next preceding layer, called a fully connected network [4]. It receives inputs, combines them, performs required computations based on a predefined activation function, and delivers the output. The devices with ANN using two terminals suffer from non-linearity and asymmetric conductance tuning problems. Due to this, their scalability was also affected. To solve these issues, an IFG (Ionic Floating Gate) model memristor-based device will be developed, resulting in a memristor-based neural network (MNN) from an ANN with memristor features. To implement this, a cadence Verilog A model will be used along with deep learning concepts. A memristor device has a simple structure with two terminals, its conductance can be modified by simple positive and negative pulses while representing synaptic weight [5]. These main functions of the memristor made it suitable for realizing the synaptic weight in an artificial neural network. By using a memristor device output as a synapse, a memristive neural network is expanded to be a multiple layer network with modified memristor-based backpropagation and gradient descent learning rules [6].

2. Memristor-Based Neural Networks

For the design of the MNN building, the network and weight loading both take a crucial role. Thereafter, the IFG model will act as a conductor. For the pattern recognition to be verified by the MNN, a dataset called MNIST is used at input neurons. Additionally, we use resistors to build a fully connected MNN. The optimization of weights is done by software [7]. The following Equation (3) is given for the output voltage.

$$V_{out} = R_{limit} . \sum_{i=1}^{\Psi} VRi . \frac{1}{R_{ij}} \tag{3}$$

Here, Rij is the reciprocal of the conductance(G) which are obtained by Cadence Virtuoso, these are further used as weights Wij. To train and test the neural network, we need a massive dataset. Fortunately, the MNIST (modified national institute of standards and technology) database exists, which contains 60,000 training images and 10,000 testing images. In this dataset, picture pixels will be normalized by greyscale numbers divided from 0 to 255 values [8].

2.1. Memristor-Based IFG Model

This Ionic Floating Gate (IFG) memristive device has three terminals and is a combination of a redox transistor and a non-volatile CBM to make it a non-volatile synaptic memory [9]. A redox transistor was developed in memristor techniques to find a way to clear limitations such as low writing efficiency, vanishing gradients, and limited accuracy. It contains three layers. The first layer is PEDOT: which is made of polystyrene sulfonate i.e., PSS film followed by the Nafion layer. The last layer is the PEI (poly(ethylenimine)) layer, which is partially reduced by PEDOT: PSS film [10]. Connected to this is a conductive bridge memory that is designed with a layer of Ag in between Pt electrodes. The IFG model can be used as a memory storage device to memorize the last operation. Memory operates in two modes: read operation and write operation. Figures 2 and 3 shows the internal view of the IFG memory [11].

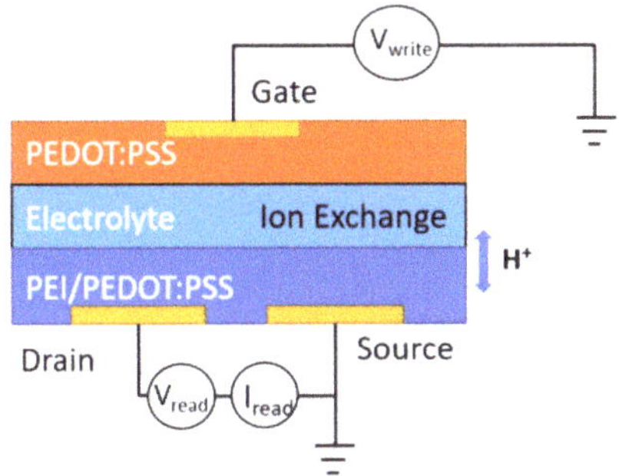

Figure 2. A polymer-based redox transistor.

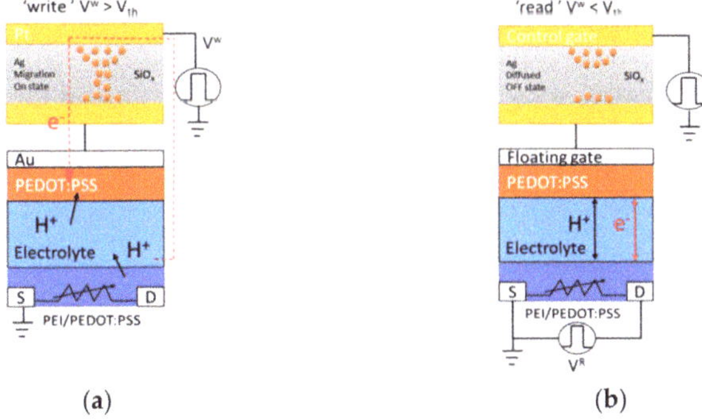

(**a**)　　　　　　　　　　(**b**)

Figure 3. Complete IFG model showing connections between redox transistor and CBM. (**a**) Write operation; (**b**) read operation.

By taking the above two models as a reference including some modifications in them, a compact IFG model characterized in Verilog-A code is developed, and importing this Verilog-A code file to cadence generates the IFG model [12]. The cadence model is shown in Figure 4. Here, we change the selector position to the electrolyte and middle of the gate terminal. These are suitable for the design of memristor-based synaptic circuits. The memristor IFG model conductance values are taken as synaptic arrays. For the recognition of patterns/images, we use a three-layer artificial network that is fully connected [13]. All weight connections hold their fixed conductance between the layers of the neural network, which is stored as IFG's memristance "G" and mapped into the source-drain conductance of the synaptic array [14]. Here, the change in conductance is proportional to the flux, its gate voltage is greater than the threshold voltage, and tunes the gate to the source voltage [15]. The schematic circuit of IFG is shown in Figure 4, which is implemented in cadence.

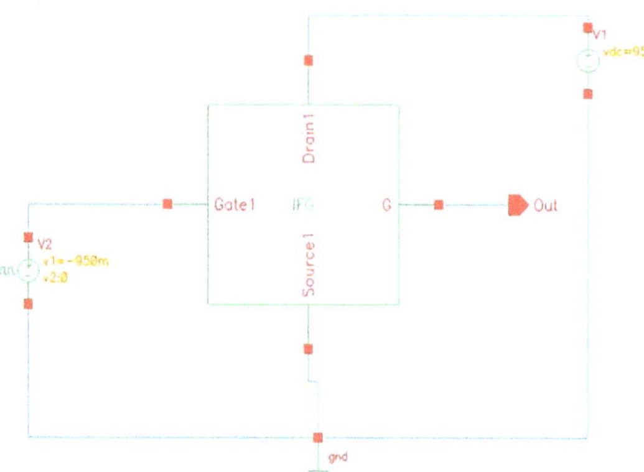

Figure 4. IFG-model-equivalent circuit. In the IFG model, the voltage applied at the "Gate" terminal is "V_w", which is the write voltage, and at the "Drain" terminal is "V_r" which is the read voltage.

The voltage applied at the gate node must be greater than the threshold voltage (V_{th}). The voltage level should be ± 0.95 V which is given to the gate and source node.

2.2. Weight Setting through Gradient Descent Model and Backpropagation

Specific learning algorithms are useful to hasten the training of neural networks. Here, we adopt the gradient descent model and backpropagation along with memristor-based IFG circuit. It takes less than a few seconds to run the ANN [16]. A text-based "Spectra" is generated for the connections in the ANN circuit with the help of python. The input for this python code is the database from cadence simulation (Synapse) [17]. First, a normal-weighted synapse is given from output of the IFG model to the neural networks after getting the optimized values from the neural networks. Again, these optimized values should be applied to IFG and observe whether our activation function is optimized. Here, the weights are optimized by using gradient descent and backpropagation models. The differences are backpropagated to the neural networks. At epoch 0.68, the accuracy increases by 0.8% as compared to the traditional method. The circuit-level design and implementation of gradient and backpropagation learning architectures help to get optimized results. These results are compared with the original one. The below given Figure 5 shows the simple operation of ANN.

To achieve image pattern recognition, the optimized values are again given as weighted synapses in the neural networks along with the images as input. Here, ReLu and Adams activation functions are used. Wij are weighted neurons, and hij is the summation of input neurons and weights. The summation is fed to the activation function, which is shown in Figure 6.

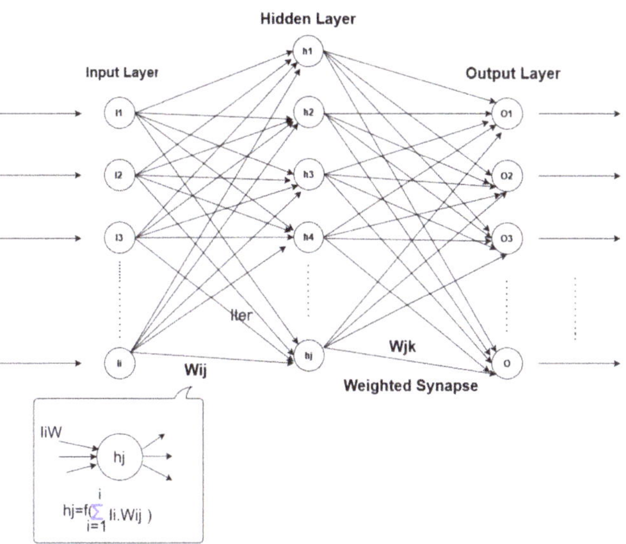

Figure 5. A simple circuit affirmation from the input layer to the output layer.

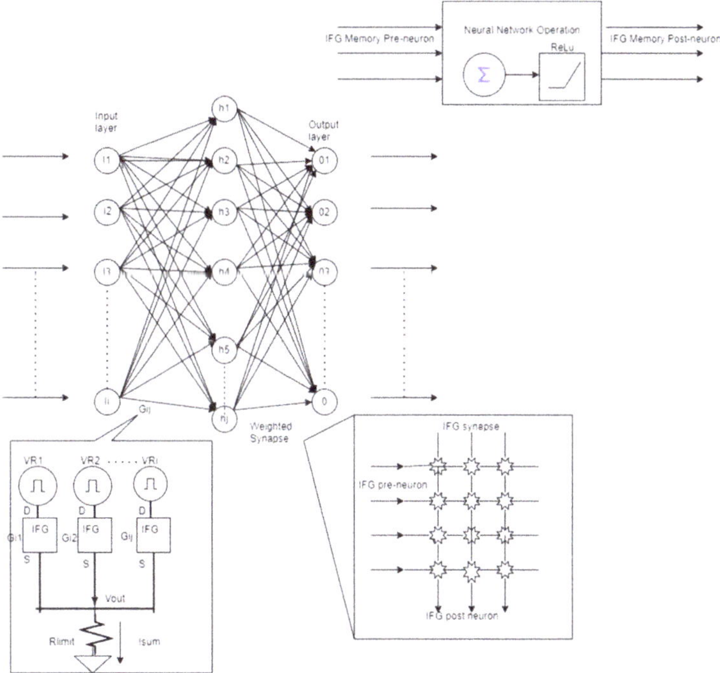

Figure 6. Operation of neural networks with IFG memory synapse and implementing image processing using gradient descent algorithm and backpropagation.

3. Results and Analysis

To check the basic memristor characteristics observe Figure 7, the sine function is given as input harmonic with voltage (V) of 1 V amplitude and 1 Hz frequency ($V(t) = V_0\sin(wt)$ for $w = 2w_0$ angular frequency where the V-I characteristics shows with the applied negative voltage the current (I) increases, and the current (I) drops with the positive voltage value, causing a hysteresis loop to emerge. We can also see the linear relationship between the charge and flux.

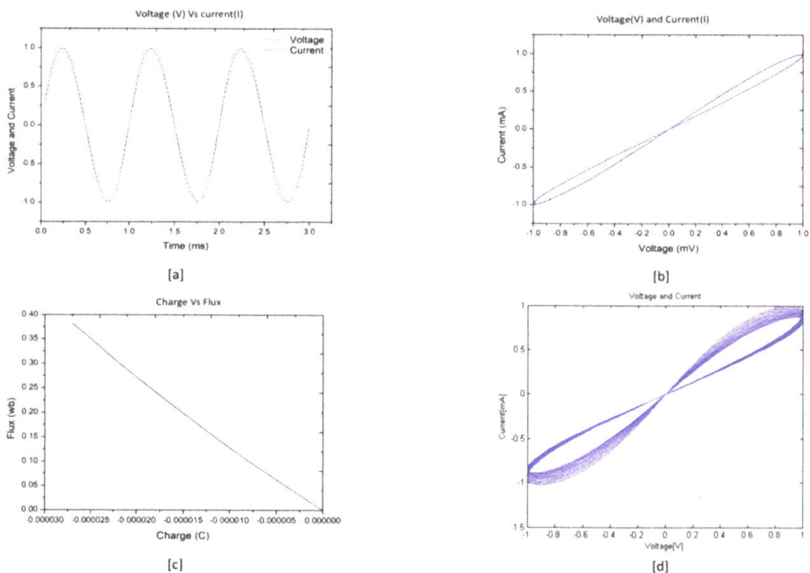

Figure 7. MATLAB simulation results of basic Memristor. (**a**) Input sinewave signal (**b**) V-I characteristics of memristor, (**c**) linear relation between charge and flux (**d**) pinched hysteresis loop (non-linear characteristics).

In the result given Figure 8, for a positive pulse, the conductance values decrease for positive input voltage and increase with the negative input voltage i.e., similar to the characteristics of a memristor. These are the simulation results of the IFG model in cadence, for which the voltage value is ±0.95 V, and the threshold voltage is 0.4 V. The input voltage is tuned in between the gate and source, the threshold voltage (V_{th}) is applied to dc source i.e., connected in between drain and source.

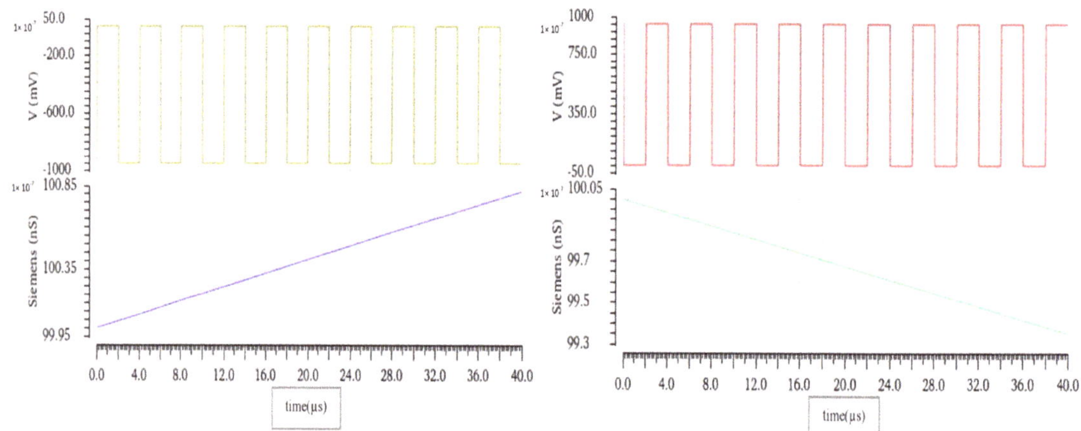

Figure 8. Two sets of graphs, one with the positive pulse response, and another one with the negative pulse response.

Figure 9 below shows the post-simulation values for the optimized values from the IFG model.

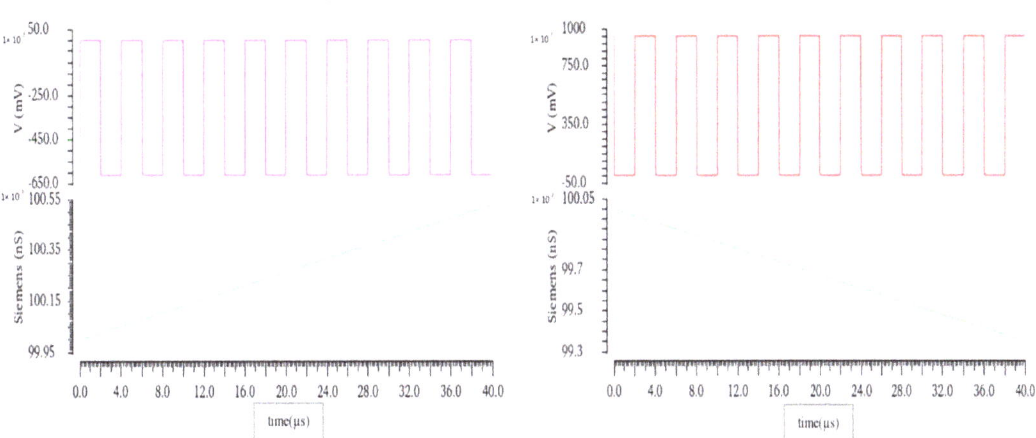

Figure 9. Post-simulation results for optimized parameters.

After applying the optimized values to the IFG device, the same operation is repeated as it was in previous case but with increased accuracy by 0.8%. Here, the optimized parameters are X = 0.150 V, and Y = −0.612 V.

Here, Table 1 mention some values of basic memristor showing how the current varies with a change in input voltage and Tables 2 and 3 mention values of IFG using Cadence Virtuoso, how conductance varies with a change in input voltage. The various parameters stated here for basic memristor are current (I), voltage (V), flux(Ψ), and charge (q). For the IFG model, the Gmax (conductance) values are given as a relationship between voltage and siemens.

Table 1. Parametric values of the current, voltage, flux and charge for hysteresis curve with reference to Figure 1.

Measurement #	Input Voltage (V)	Input Current (I)	Flux (Ψ)	Charge (q)
1.	0.018 mV	0.022 mA	3.3×10^{-9} (Wb)	-2.8×10^{-9} C
2.	0.037 mV	0.044 mA	1.3×10^{-4} (Wb)	-1.1×10^{-8} C
3.	0.056 mV	0.066 mA	1.5×10^{-4} (Wb)	-2.5×10^{-8} C
4.	0.075 mV	0.088 mA	1.8×10^{-4} (Wb)	-4.5×10^{-8} C

Table 2. Parametric value analysis when the input pulse response = (+0.95 V).

Measurement #	Input Voltage (V)	Siemens (ns)	
		Gmax = 1×10^7(X)	Gmax = 1×10^7(Y)
1.	0.95 V	0	1.0×10^{-7}
2.	0.60 V	2.0×10^{-8}	1.0×10^{-7}
3.	0.31 V	3.0×10^{-7}	1.0×10^{-7}
4.	0.19 V	4.02×10^{-6}	9.9×10^{-8}
5.	0	6.02×10^{-6}	9.9×10^{-8}

Table 3. Parametric value analysis when the input pulse response = (-0.95 V).

Measurement #	Input Voltage (V)	Siemens (ns)	
		Gmax = 1×10^{-7}(X)	Gmax=1×10^{-7}(Y)
1.	0	0	1.0×10^{-7}
2.	-0.19 V	2.0×10^{-8}	1.0×10^{-7}
3.	-0.31 V	3.0×10^{-7}	1.0×10^{-7}
4.	-0.60 V	4.0×10^{-6}	1.0×10^{-7}
5.	-0.95 V	6.0×10^{-6}	1.0×10^{-7}

By observing Table 4, the traditional model is compared with proposed model, in traditional model sigmoid function is used and in proposed model ReLu and Adams functions are used as activation function in neural network with optimized parameters causes the accuracy of recognition to increased up to 94.6%. So, with the proposed IFG model, the accuracy increases.

Table 4. Comparison between analytical values and optimized values.

Model #	Activation Function	Accuracy
Traditional model	Sigmoid function	93.8%
Proposed model	ReLu function and Adams function	94.6%

4. Conclusions

Artificial neural networks have a vital role in unsupervised deep learning models to implement applications based on pattern recognition and also in many areas of our daily lives. This paper provides us with the knowledge of developing an IFG device using cadence with the characteristics of a memristor-based circuit. To demonstrate the capability of an IFG device in pattern recognition, the values are optimized using the gradient descent model and the resulting optimized parameters are compared with the existing parameters. The testing ability of neural network increased by 0.8% when using the IFG model. Hence, the accuracy of the original network is 93.8%, which increases to 94.6%. Here, ReLu and Adams activation helps in faster optimization, and by this result, we can conclude that, by using IFG-based memristor characteristics in neural networks, one can increase image recognition accuracy.

Author Contributions: G.R.: Running experiments and drafting paper. N.P.S.: Supervision and proofreading. S.R.: Running experiments and analysis. E.S.K.R.: Running experiments and drafting paper. All authors have read and agreed to the published version of the manuscript.

Funding: This research received no external funding.

Institutional Review Board Statement: Not applicable.

Informed Consent Statement: Not applicable.

Data Availability Statement: Data can be shared on request.

Conflicts of Interest: The authors declare no conflict of interest.

References

1. Creswell. UC Santa Cruz UC Santa Cruz Electronic Theses and Dissertations Title. 2020. Available online: https://escholarship.org/uc/item/0jx2107r (accessed on 1 June 2020).
2. Amelia, D. Application of Memristive Device Arrays for Pattern Recognition-Title. 2020. Available online: http://mpoc.org.my/malaysian-palm-oil-industry/ (accessed on 1 June 2020).
3. Adhikari, S.P.; Kim, H.; Budhathoki, R.K.; Yang, C.; Chua, L.O. A circuit-based learning architecture for multilayer neural networks with memristor bridge synapses. *IEEE Trans. Circuits Syst. I Regul. Pap.* **2015**, *62*, 215–223. [CrossRef]

4. Krestinskaya, O.; Salama, K.N.; James, A.P. Learning in memristive neural network architectures using analog backpropagation circuits. *IEEE Trans. Circuits Syst. I Regul. Pap.* **2019**, *66*, 719–732. [CrossRef]
5. Krestinskaya, O.; Ibrayev, T.; James, A.P. Hierarchical Temporal Memory Features with Memristor Logic Circuits for Pattern Recognition. *IEEE Trans. Comput. Des. Integr. Circuits Syst.* **2018**, *37*, 1143–1156. [CrossRef]
6. Chu, M.; Kim, B.; Park, S.; Hwang, H.; Jeon, M.; Lee, B.H.; Lee, B.G. Neuromorphic Hardware System for Visual Pattern Recognition with Memristor Array and CMOS Neuron. *IEEE Trans. Ind. Electron.* **2015**, *62*, 2410–2419. [CrossRef]
7. Zhang, Y.; Li, Y.; Wang, X.; Friedman, E.G. Synaptic Characteristics of Ag/AgInSbTe/Ta-Based Memristor for Pattern Recognition Applications. *IEEE Trans. Electron Devices* **2017**, *64*, 1806–1811. [CrossRef]
8. Yan, R.; Hong, Q.; Wang, C.; Sun, J.; Li, Y. Multilayer Memristive Neural Network Circuit Based on Online Learning for License Plate Detection. *IEEE Trans. Comput. Des. Integr. Circuits Syst.* **2022**, *41*, 3000–3011. [CrossRef]
9. Abdoli, B.; Amirsoleimani, A.; Shamsi, J.; Mohammadi, K.; Ahmadi, A. A Novel CMOS-Memristor Based Inverter Circuit Design. In Proceedings of the 2014 22nd Iranian Conference on Electrical Engineering (ICEE), Tehran, Iran, 20–22 May 2014. [CrossRef]
10. Azghadi, M.R.; Linares-Barranco, B.; Abbott, D.; Leong, P.H.W. A Hybrid CMOS-Memristor Neuromorphic Synapse. *IEEE Trans. Biomed. Circuits Syst.* **2017**, *11*, 434–445. [CrossRef] [PubMed]
11. Querlioz, D.; Bichler, O.; Dollfus, P.; Gamrat, C. Immunity to device variations in a spiking neural network with memristive nanodevices. *IEEE Trans. Nanotechnol.* **2013**, *12*, 288–295. [CrossRef]
12. Ran, H.; Wen, S.; Li, Q.; Yang, Y.; Shi, K.; Feng, Y.; Zhou, P.; Huang, T. Memristor-Based Edge Computing of Blaze Block for Image Recognition. *IEEE Trans. Neural Netw. Learn. Syst.* **2022**, *33*, 2121–2131. [CrossRef]
13. Xu, X.; Xu, W.; Wei, B.; Hu, F. Memristor-based neural network circuit of delay and simultaneous conditioning. *IEEE Access* **2021**, *9*, 148933–148947. [CrossRef]
14. Zhang, Y.; Wang, X.; Li, Y.; Friedman, E.G. Memristive Model for Synaptic Circuits. *IEEE Trans. Circuits Syst. II Express Briefs* **2017**, *64*, 767–771. [CrossRef]
15. Duan, Q.; Jing, Z.; Zou, X.; Wang, Y.; Yang, K.; Zhang, T.; Wu, S.; Huang, R.; Yang, Y. Spiking neurons with spatiotemporal dynamics and gain modulation for monolithically integrated memristive neural networks. *Nat. Commun.* **2020**, *11*, 3399. [CrossRef]
16. Sacchetto, D.; Gaillardon, P.E.; Zervas, M.; Carrara, S.; De Micheli, G.; Leblebici, Y. Applications of multi-terminal memristive devices: A review. *IEEE Circuits Syst. Mag.* **2013**, *13*, 23–41. [CrossRef]
17. Truong, S.N. Single Crossbar Array of Memristors with Bipolar Inputs for Neuromorphic Image Recognition. *IEEE Access* **2020**, *8*, 69327–69332. [CrossRef]

Proceeding Paper

Digital Polarization Holography: Challenges and Opportunities †

Rakesh Kumar Singh

Department of Physics, Indian Institute of Technology (Banaras Hindu University),
Varanasi 221005, Uttar Pradesh, India; krakeshsingh.phy@iitbhu.ac.in
† Presented at the International Conference on "Holography Meets Advanced Manufacturing", Online,
  20–22 February 2023.

Abstract: Polarization has a profound impact on image quality and visual perception. For instance, polarization provides a new perspective on seeing an object which is otherwise obscured, low contrast or not measurable by conventional imaging methods. In this paper, we discuss a possible extension of the digital holography (DH) to the polarization domain, and the technique is referred to as digital polarization holography (DPH). The basic principle of the DPH is described and some of our recent contributions on quantitative vectorial imaging are covered. We also discuss and highlight the potential of combining speckle field illumination with DPH for high-resolution vectorial imaging.

Keywords: digital holography; polarization; imaging

Citation: Singh, R.K. Digital Polarization Holography: Challenges and Opportunities. *Eng. Proc.* **2023**, *34*, 10. https://doi.org/10.3390/HMAM2-14112

Academic Editor: Vijayakumar Anand

Published: 4 March 2023

1. Introduction

Holography uses the principle of interference to record and reconstruct complex valued objects. The availability of high-quality detectors and computational facilities has popularized holography through digital holography (DH), where a hologram is digitally recorded and a reconstruction of the hologram is implemented by numerical methods [1]. DH provides a quantitative information of the complex fields, i.e., amplitude and phase. This is advantageous in the context of real-time live quantitative imaging, digital depth focusing, 3D, and label-free nondestructive imaging. The availability of array detectors and reconstruction algorithms have further revolutionized holography. Different DH schemes have been developed, and significant among them are in-line, off-axis, and phase-shifting holography. Off-axis holography requires angularly separated object and reference beams in an interferometric design, and this geometry avoids the twin image problem in the reconstruction. A concept of preserving information in the interference fringes has also been tested for self-luminous or incoherent objects. Significant techniques among the incoherent holography are coherence holography [2] and Fresnel incoherent correlation holography (FINCH) [3]. Coherence holography is an unconventional approach, wherein the complex valued object is reconstructed as a spatial distribution of the complex coherence function. In another development, DH methods are used to image 2D and 3D objects located behind the random scattering medium [4–6].

DH has emerged as a unique technique to quantitatively measure the wavefront of light. A DH combined with microscopy, called digital holographic microscopy (DHM) is nowadays widely used for a wide range of applications [1]. However, a complete description of the wavefront needs the inclusion of the polarization vector in its measurement and analysis [7]. Polarization analysis is important in fields such as stress analysis, bio-medical imaging, chemistry and so on. Polarization states, a significant parameter to describe light matter interactions, have been critical and significant in contrast enhancement, and highlight specific cell structures which are otherwise missing in the scalar imaging. Therefore, polarization imaging is considered to be a promising and futuristic tool, as it is capable of revealing the order at a molecular scale that is usually hidden to the conventional microscopes. Lohman was the first to propose the issue of the total recording and reconstruction of the wavefront by including the polarization vector in the wavefront reconstruction [7]. The extension of holography to the polarization domain is possible by recording and reconstructing two holograms of the orthogonal polarization components [8–12]. In order to expand polarization holography to the light with arbitrary coherence, it is desirable to examine the interference using the Stokes parameters of the light [9,10]. A practical challenge in polarization holography is the recording and reconstruction of four holograms which correspond to all four Stokes parameters. This challenge has been attempted by using random scattering as a real-time recording plane and reconstructing the polarization vector by two-point complex correlations of the random light field. Polarization imaging of three-dimensional imaging using phase-shifting holography is also possible.

In this paper, we discuss polarization interference fringes and their role in the recording and reconstruction of the complete wavefront. To demonstrate the usefulness of the polarization fringes in the quantitative and spatially resolved imaging, beyond conventional DH, we perform the recording and reconstruction of the polarization holograms with a generic light source of various correlations. We discuss different experimental designs of digital polarization holography (DPH) and some of our recent contributions on quantitative vectorial imaging. A new approach in the polarization domain of holograph, called speckle-field digital polarization holographic microscopy (SDPHM) is also discussed. The idea is to use a random pattern (rather than a uniform field) for illumination of the sample and thereby recover the high-resolution polarization features in comparison to the scalar DPH.

2. Polarization Holograms

2.1. Digital Polarization Holography

Consider the recording geometry represented in Figure 1. This figure shows the recording of a hologram by interference of the waves emerging from two sources, i.e., the object and reference. The vectorial nature of the light waves, on the plane and vertical to the plane, are represented by a green arrow and black circle. Here, the arrow represents the polarization vector on the plane and black circle represents a polarization vector perpendicular to this plane. The propagation of the light field from the source to the observation is represented as

$$O_s(r) = \int O_s(\rho)G(r - \rho)d\rho \tag{1}$$

$$R_s(r) = exp[i\alpha r] \tag{2}$$

where $O_s(r)$ and $O_s(\rho)$ represent a realization of the object field at the observation and source plane, respectively, and $s = x, y$ represents the orthogonal polarization components; and $G(r - \rho)$ is the propagation kernel to accommodate diffraction from the source position (ρ) to the spatial position (r) at the observation plane. A reference beam $R_s(r)$ is considered to be uniform with a linear phase of spatial frequency α.

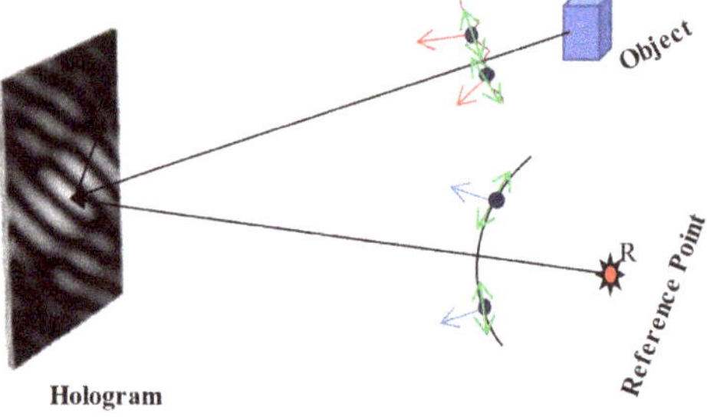

Figure 1. A schematic diagram to represent recording of the hologram.

2.2. Recording and Reconstruction of the Polarization Hologram

A single realization of orthogonally polarized components, at the observation plane, are represented as

$$E_x(r) = O_x(r) + R(r) \tag{3}$$

$$E_y(r) = O_y(r) + R(r) \tag{4}$$

Stokes fringes can be obtained from the orthogonal polarization components as

$$S_o(r) = \langle E_x^*(r)E_x(r) \rangle + \left\langle E_y^*(r)E_y(r) \right\rangle \tag{5}$$

$$S_1(r) = \langle E_x^*(r)E_x(r) \rangle - \left\langle E_y^*(r)E_y(r) \right\rangle \tag{6}$$

$$S_2(r) = \langle E_x^*(r)E_y(r) \rangle + \left\langle E_y^*(r)E_x(r) \right\rangle \tag{7}$$

$$S_3(r) = i\left[\left\langle E_y^*(r)E_x(r) \right\rangle - \langle E_x^*(r)E_y(r) \rangle \right] \tag{8}$$

where $<.>$ angular bracket represent the ensemble average to evaluate the statistical properties of the light. Four Stokes parameters encode coherence-polarization features of the sources at the recording plane. The first Stokes parameter in Equation (5), i.e., $S_o(r)$ represents the intensity and corresponds to the conventional hologram recording. Either a combination of $S_o(r)$ and $S_1(r)$ or $S_2(r)$ and $S_3(r)$ is sufficient to reconstruct the full field of the object in a coherent recording. A detailed discussion on recording and reconstruction of the Stokes hologram can be found in Refs. [9,10].

3. Experimental Design and Implementation

In order to consider the recording and reconstruction of holograms from the coherent and fully polarized source, we confine it to the first two Stokes parameters and use holograms of the orthogonally polarized components, i.e., $I_x(r) = |E_x(r)|^2$ and $I_y(r) = |E_y(r)|^2$. An experimental scheme to simultaneously record these two holograms are shown in Figure 2. A specially designed Mach–Zehnder type polarization interferometer equipped with a triangular Sagnac geometry is used to simultaneously record the orthogonal polarization components. A collimated diagonally polarized coherent beam is split into two copies by a beam splitter (BS1). A beam, transmitted by BS1 and folded by the mirror M1, illuminates the sample. This object beam is imaged at the camera plane through the BS2. On the other hand, a beam reflected by the BS1 works enters into a triangular Sagnac interferometer assisted with a telescopic lens assembly (L2 and L3) to generate a distinguishable orthogonally polarized reference beam. A polarization beam splitter (PBS) splits into counter propagating orthogonally polarized components. The mirrors M2 and M2 introduce spatial separation in the orthogonally polarized components at the back focal plane of lens L2 which is also a front focal plane of lens L3. Thus, the orthogonally polarized components, coming out of the triangular Sagnac geometry, gain distinguishable linear phases at the back focal plane of lens L3 which overlaps with the detector plane. The angular multiplexed reference beams is a significant feature of our experimental design and has been utilized for single-shot polarization imaging [11] and Jones matrix microscopy [13]. These reference beams interfere with the object beam and a hologram is recorded by the charge coupled device (CCD). The intensity, recorded by the CCD, is represented as

$$I(r) = |O_x(r) + R_x(r)|^2 + |O_y(r) + R_y(r)|^2$$

where $R_x(r)$ and $R_y(r)$ are reference beams for the x and y polarization components, respectively. Orthogonal polarization components of the object field are represented as $O_x(r) = |O_x(r)|e^{i\varphi_x(r)}$ and $O_y(r) = |O_y(r)|e^{i\varphi_y(r)}$.

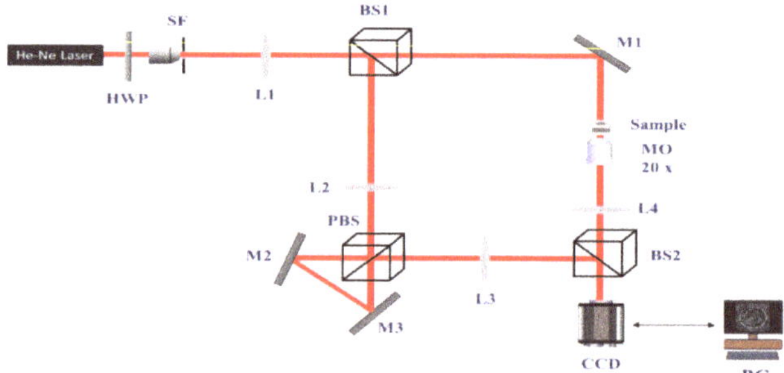

Figure 2. An experimental setup to record polarization hologram.

For demonstration purposes, we present an interference pattern of the birefringence object in Figure 2. A mesh structure in the interference pattern, as highlighted in the inset, appears due to distinguishable carrier frequencies of the orthogonally polarized reference beams. This scheme is very important for recording and reconstructing the spatially varying polarization of the object from a single hologram recording [11]. This recorded hologram is subjected to digital Fourier fringe analysis to reconstruct the complex fields of the orthogonally polarized components, and results are shown on the right hand side in Figure 3.

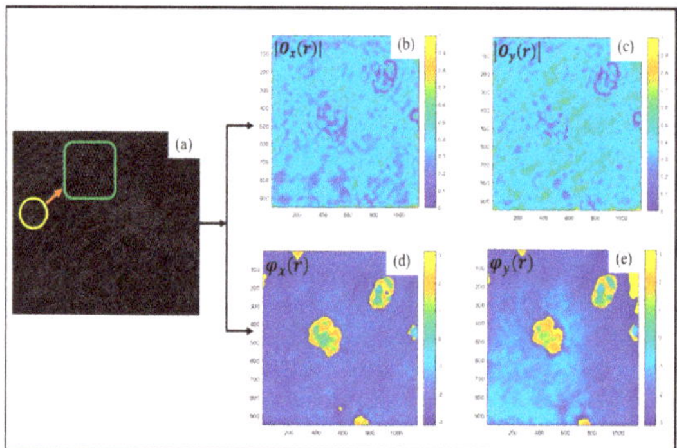

Figure 3. Reconstruction of complex fields of the orthogonal polarization components from a recorded hologram. (**a**) represents the digitally recorded hologram. (**b**,**c**) represent amplitude distributions of the orthogonal polarization components. Corresponding phase distributions are shown in (**d**,**e**).

4. Challenges and Opportunities

In general, all four Stokes fringes are desired for a generic imaging application. An intensity fringe is one of the four Stokes fringes. Thus, the recording and reconstruction of all four fringes may require multiple measurements by varying the polarization optics or using the polarization sensitive recording medium. A challenge of multiple-step recording can be addressed by using a polarization-sensitive detector or using the random scattering medium as a real-time recording medium with correlation-based reconstruction. The importance of the Stokes fringes in polarization imaging can be highlighted by the example of interference between the orthogonal polarization components which is not possible to observe using conventional digital holography. Considering an off-axis object with a horizontally polarized transmittance, i.e., $O_x(\rho) = |O_x(\rho)|e^{i\varphi_x(\rho)}$, and reference with a vertical polarization, i.e., $R_y(\rho) = 1$, the first two Stokes parameters at the recording plane, i.e., $S_0(r)$ and $S_1(r)$, are made of only non-modulating terms and no interreference appears in these fringes as expected from the scalar theory. On the other hand, the Stokes fringes $S_2(r)$ and $S_3(r)$ preserve a spatial carrier frequency introduced by the cross-modulations between the off-axis object and the on-axis reference source [10]. The recording and digital reconstruction of the polarization holograms have also offered a new opportunity to use speckle field illumination for high-resolution polarization imaging, and the technique is referred as speckle-field digital polarization holographic microscopy (SDPHM) [13]. A digitally generated speckle pattern illumination in the SDPHM offers a new opportunity to explore high resolution polarization imaging.

5. Conclusions

The possible extension of digital holography to the polarization domain is discussed and some of our recent contributions are briefly discussed in this paper. A special emphasis is given to the recording and reconstruction of complete and spatially resolved polarimetric features from a single hologram.

Funding: This work is supported by the Science and Engineering Research Board (SERB) India CORE/2019/000026 and the Department of Biotechnology (DBT) Grant No. BT/PR35587/MED/32/707/2019.

Institutional Review Board Statement: Not applicable.

Informed Consent Statement: Not applicable.

Data Availability Statement: Not applicable.

Acknowledgments: This paper is based on research contributions of former and present lab members. R. K. Singh acknowledges support from the IIT (BHU) and different funding agencies.

Conflicts of Interest: The author declares no conflict of interest.

References

1. Schnars, U.; Jueptner, W. *Digital Holography*, 1st ed.; Springer: Berlin/Heidelberg, Germany, 2005.
2. Takeda, M.; Wang, W.; Duan, Z.; Miyamoto, Y. Coherence holography. *Opt. Express* **2005**, *13*, 9629–9635. [CrossRef] [PubMed]
3. Rosen, J.; Brooker, G. Digital spatially incoherent Fresnel holography. *Opt. Lett.* **2007**, *32*, 912–914. [CrossRef] [PubMed]
4. Naik, D.N.; Singh, R.K.; Ezawa, T.; Miyamoto, Y.; Takeda, M. Photon correlation holography. *Opt. Express* **2010**, *19*, 1408–1421. [CrossRef] [PubMed]
5. Singh, R.K.; Vinu, R.V.; Sharma, A. Recovery of complex valued objects from two-point intensity correlation measurement. *Appl. Phys. Lett.* **2014**, *104*, 111108. [CrossRef]
6. Singh, R.K.; Sharma, A.M.; Das, B. Quantitative phase contrast imaging through a scattering medium. *Opt. Lett.* **2014**, *39*, 5045–5057. [CrossRef] [PubMed]
7. Lohmann, A.W. Reconstruction of vectorial wavefronts. *Appl. Opt.* **1965**, *4*, 1667–1668. [CrossRef]
8. Colomb, T.; Dahlgren, P.; Beghuin, D.; Cuche, E.; Marquet, P.; Depeursinge, C. Polarization imaging by use of digital holography. *Appl. Opt.* **2002**, *41*, 27–37. [CrossRef] [PubMed]
9. Singh, R.K.; Naik, D.N.; Itou, H.; Miyamoto, Y.; Takeda, M. Stokes holography for recording and reconstructing objects using polarization fringes. *Proc. SPIE* **2011**, *8082*, 808209.
10. Singh, R.K.; Naik, D.N.; Itou, H.; Miyamoto, Y.; Takeda, M. Stokes holography. *Opt. Lett.* **2012**, *37*, 966–996. [CrossRef] [PubMed]
11. Sreelal, M.M.; Vinu, R.V.; Singh, R.K. Jones matrix microscopy from a single-shot intensity measurement. *Opt. Lett.* **2017**, *42*, 5194–5197. [CrossRef] [PubMed]
12. Singh, D.; Singh, R.K. Lensless Stokes holography with the Hanbury Brown-Twiss approach. *Opt. Express* **2018**, *26*, 10801–10812. [CrossRef] [PubMed]
13. Vinu, R.V.; Chen, Z.; Pu, J.; Otani, Y.; Singh, R.K. Speckle-field polarization holographic microscopy. *Opt. Lett.* **2019**, *44*, 5711–5714. [CrossRef] [PubMed]

Proceeding Paper

Field of View Enhancement of Dynamic Holographic Displays Using Algorithms, Devices, and Systems: A Review †

Monika Rani, Narmada Joshi, Bhargab Das and Raj Kumar *

CSIR—Central Scientific Instruments Organization, Sector 30C, Chandigarh 160030, India; monikarani@csio.res.in (M.R.); narmadajoshi1994@gmail.com (N.J.); bhargab.das@csio.res.in (B.D.)
* Correspondence: raj.optics@csio.res.in
† Presented at the International Conference on "Holography Meets Advanced Manufacturing", Online, 20–22 February 2023.

Abstract: Holography is a prominent 3D display approach as it offers a realistic 3D display without the need for special glasses. Due to advancements in computation power and optoelectronic technology, holographic displays have emerged as widely appreciated technology among other 3D display technologies and have drawn a lot of research interest in recent years. The core of dynamic holographic displays is spatial light modulator (SLM) technology. However, owing to the limited resolution and large pixel size of SLMs, holographic displays suffer from certain bottlenecks such as limited field of view (FOV) and narrow viewing angle. To develop a holographic display at the commercial level, it is crucial to solve these problems. A variety of probable solutions to these challenges may be found in the literature. In this review, we discuss the essence of these approaches. We study the important milestones of the various methodologies from three primary perspectives— algorithms, optical systems, and devices employed for FOV extension—and provide useful insights for future research.

Keywords: holographic displays; field of view; spatial light modulator

1. Introduction

In order to make their thoughts come to life, people want to visualize them. The methods through which thoughts and experiences are visualized have evolved from art to photography and more recently to 3D displays with the advancements of science and technology. The four main types of 3D displays are stereoscopic, auto-stereoscopic, volumetric, and holographic displays [1]. Holography has been recognized as the most suitable method for displaying 3D images that seem realistic. It has the ability to control crucial aspects of light, such as phase, which cannot be controlled by other methods [2]. Holographic three-dimensional (3D) displays, which can rebuild 3D images with complete depth cues, have obtained broad attention in last few decades for their potential applications in a variety of industries, including the medical and military sectors [3]. Holographic dynamic displays come into picture in order to display information at video rate. Since video-enabled holography needs a lot of pixels and data, implementing it is another difficult task. However, evolution in parallel computing approaches and inventions of optical modulators make the dynamic loading as well as dynamic display of holograms possible. To display the calculated digital holograms, spatial light modulators (SLM) are one of the widely used methods [4]. However, field of view (FOV), space bandwidth product (SBP) of the hologram, and the display quality are all constrained by the size and pixel pitch of commercially available SLMs. As a result, the SBP of holographic displays utilizing SLM is often several hundred times lower than that of static holographic media. It implies that for an SLM to display a hologram, either the hologram size or the viewing angle must be inadequate. To achieve a large sized, dynamic, full-color, holographic 3D display, it is important to overcome these problems. So, in the past few years, there have been various attempts to

Citation: Rani, M.; Joshi, N.; Das, B.; Kumar, R. Field of View Enhancement of Dynamic Holographic Displays Using Algorithms, Devices, and Systems: A Review. *Eng. Proc.* **2023**, *34*, 11. https://doi.org/10.3390/ HMAM2-14129

Academic Editor: Vijayakumar Anand

Published: 7 March 2023

overcome these barriers for realizing a commercially viable dynamic holographic display technology.

In this review study, we take a look at some of the prominent methods that are proposed by different researchers to overcome the problems associated with limited FOV of holographic displays. This study is divided into three sections. In Section 1, the FOV expansion approaches based on algorithms used for CGH generation are discussed. The introduction and principle of various systems and devices employed in optical configurations are explained in Sections 2 and 3, respectively. The advantages and disadvantages regarding each method are also explored from the realization/employment perspective. From this study, we investigate what more is required to have a holographic display with wide FOV in our hands.

2. Algorithmic Approaches for CGH Generation

One of the most important aspects of holographic displays is computer generated hologram (CGH) generation. The objective of CGH is to model, calculate, and encode holograms from 3D scenes. Algorithms are often separated into three paths from a computational perspective, depending on the various solutions to the wave equation. The approaches mentioned in the literature include point source, polygon, spherical, and cylindrical [5]. The main objective of these algorithms is the real-time computation of complex holographic patterns with huge information capacities for multi-color, wide-angle, and large-image systems. However, the spatial frequency and physical size of the CGH affect its information capacity and FOV. As a result, a significant amount of study on overcoming these limitations is reported. Generally, curved holograms such as computer-generated cylindrical holograms (CCGH) are thought to be the best for enlarging FOV as they offer the ability to view an image from any angle, enabling the reconstruction and 360° horizontal viewing of the object [6]. For a large compensation distance, it is impossible to correctly perform the point-to-point phase compensation between wavefront recording plane and CCGH. The approximate phase compensation (APC) technique also has this restriction condition. As a result, a full CCGH cannot be produced, and the FOV extension is constrained, making it impossible to fully use the 360° FOV of the CCGH. This issue has been resolved by gapless splicing of multi-segment cylindrical holograms (mSCH) [7]. In this suggested approach, the limited condition of APC technique was initially examined. Further, the crucial compensation distance requirement was also mentioned. Three SCHs were taken into account as a group in this method and, two by two, were spliced together at a time. To obtain wide FOV display with the above-discussed method, curved display screen and flexible display materials are required. It is often convenient to employ LCoS SLMs with inclined illumination since they operate in reflection. The inclination geometry changes the FOV of the holographic display. Modifying an object wavefront obtained from recorded digital holograms for reconstruction at the tilted SLM display is a major challenge with this technology. Kozacki [8] developed a digital hologram processing technique that uses just the required paraxial holographic field by rigorously propagating paraxial fields between inclined planes. Other significant considerations in this method are pixel response and wavefront aberration calibration, SLM calibration based on tilt value, and characterization of tilt-dependent imaging space. Unconventional angular multiplexing techniques [9] are also utilized to multiplex the whole object information in single CGH according to the SLM parameters. This approach results in angle multiplexed CGHs as shown in Figure 1. Although this method only requires one SLM, the FOV it produces is insufficient for binocular vision. Another CGH generation approach [10] is developed based on angular multiplexing, but it utilizes three SLMs in a planar configuration to reconstruct the image. So, it is economically inefficient, and optical configuration also becomes complex due to the use of three SLMs. The two-step Fresnel diffraction method [11] for CGH calculation is formulated for expansion of FOV without any physical change in the optical setup. The virtual image size is increased, resulting in enhanced FOV, by decreasing the sampling interval on the intermediate plane. However, the proposed approach shows promise for the construction of holographic AR near-eye displays with wide FOV. Thus, it is clear from the literature that the algorithms used for CGH generation play a vital role in FOV expansion for a particular holographic display.

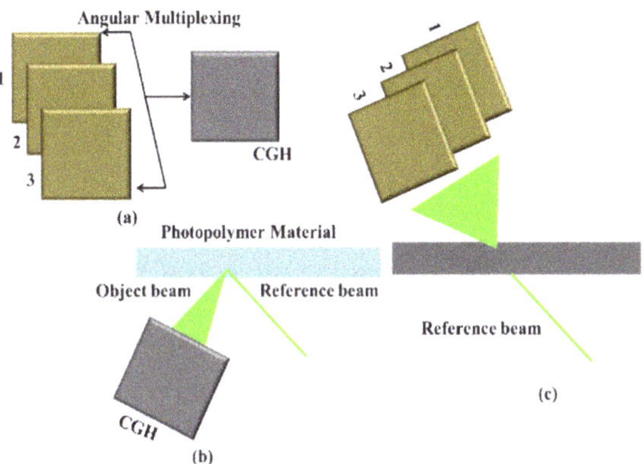

Figure 1. (a) Angular multiplexing of different objects in single CGH, (b) recording, and (c) reconstruction.

3. Configuration of Display Systems

SLMs are used in holographic displays to replicate the wavefronts of an object. Due to employment of SLMs as display systems, the FOV of reconstructed images is relatively constrained because of the small size of available SLMs. It is difficult for the viewer to move around while watching the reconstructions since the viewing angle is also small due to the large pixel pitch of SLMs. Thus, the depth cue provided by motion parallax is insufficient. Additionally, the size of the reconstructed object is rather modest due to the restricted FOV [12]. To increase the field of view for holographic displays, several techniques are reviewed in this section. As the SLMs can work in reflection mode, it is easy to illuminate them by tilted plane wave. The FOV of the display increases asymmetrically with respect to the tilt angle [8]. The performance of SLM is impacted by incidence angle. In this approach, the limitation is that digital holograms cannot be directly reconstructed due to the geometry and large tilt angles affecting the quality of the reconstructed images. Another display system is reported by Maeno et al. [13], in which five transmission type liquid crystal SLMs are placed horizontally side by side. Consequently, there is an expansion of the horizontal FOV. In addition, this technique has limited vertical parallax. To answer this problem, one more method is proposed to enhance the FOV of the reconstructed images. The field of view significantly grows along with the number of SLMs [14] as shown in Figure 2. In the reported geometry, six phase-only SLMs are used in a holographic display system that creates holographic reconstructions from a point cloud that is taken from a three-dimensional object [5]. These SLMs are tiled in a three by two pattern. However, this method is very expensive because the field of view relies on the number of modulators being employed. Further, Hahn et al. explained another unique strategy [15], in which the display system is made up of a curved array of SLMs. The spatial bandwidth of SLMs is reduced by the curved arrays, which produce more data points. The local angular spectra of the object wave are shown by individually modifying each SLM in the curved array. This configuration has a significant impact on optically reconstructed holographic images.

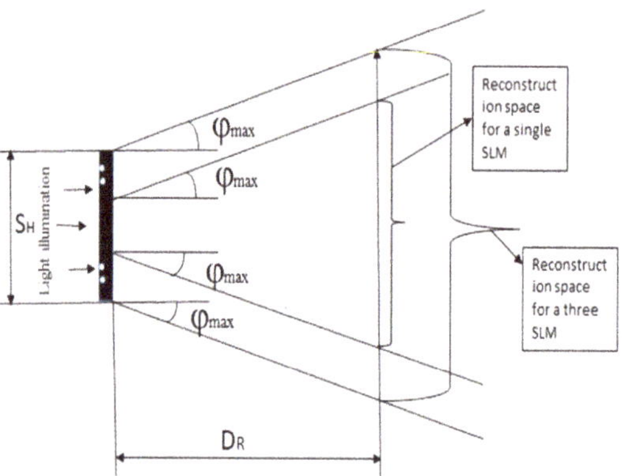

Figure 2. Reconstruction space increased with multiple SLMs.

4. Development of Optical Devices

In holographic display systems, after CGH generation and display systems, optical configurations are realized for spatial filtering of higher diffraction orders and magnification/demagnification, propagating the reconstructed image to the desired distance at the observation plane. The optical devices used in the optical configurations also have a significant contribution in the enhancement of FOV. To increase the FOV by time division and space division, multiplexing of SLMs, multiple projection systems, and CGH generation at high frame rate are required. Due to the use of a number of SLMs, there are gaps observed in the reconstructed image that degrades its quality. A holographic function screen [16] is developed to remove these gaps and further expand the FOV of display systems. This screen has a specific diffusion angle related with the diffraction angle of SLM, and it is placed at the image plane. Three SLMs are used, and the FOV is increased by 38 times as compared to single SLM with gapless splicing using this functional screen. The system becomes complex, and the functional screen reduces the intensity of light reached at the observation plane. Holographic optical elements (HOE) [17] are suggested to achieve the 80° FOV for round view. CGH is uploaded using a phase-only SLM, and noise produced by the dead zone of the SLM is blocked using a 4f optical system and spatial filtering at the Fourier plane. The HOE serves as an eyepiece and, as illustrated in Figure 3, converges a highly off-axis beam into an on-axis beam in front of the human eye. The FOV expansion is dependent upon the photopolymer material used for recording as well as the grating structure of HOE. Gu et al. [18], obtained a diagonal FOV about 55° using polarized volume holographic grating (PVG). The angular bandwidth and refractive index modulation of recorded gratings are the responsible factors for the limited FOV. In spite of the fact that PVG has a wider angular bandwidth than volume holographic grating (VHG), a single PVG is unable to accommodate a broad field of view. Thus, the laminated composite PVGs are proposed in this work. The lens array 4f system [19] is adopted to increase the FOV from 1.9° to 7.6° and optimize the SLM phase profile to improve the reconstructed image quality. Kim et al. [20] suggested a FOV expansion technique for the holographic near-eye display without additional mechanical device and micro structured mask. The original intensity is divided into pieces, and each serves as a target profile for each depth plane. The placement of beam splitter array and eyepiece lens with high numerical aperture allows realization of the enlarged FOV. The three times enlarged FOV is achieved by the mentioned approach. Based on the above research studies, it is observed that the development of such devices is helpful in the realization of holographic displays with wide FOV.

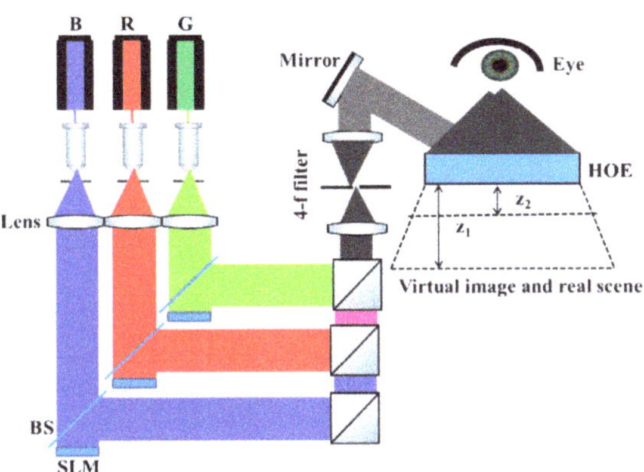

Figure 3. Schematic View of wide field see-through holographic display.

5. Conclusions

This paper reviews the current state-of-the-art FOV expansion for the holographic displays. CGH generation algorithms, configuration of display systems, and the optical devices are identified as the three most promising candidates for the above purpose. The combination of different primitive methods into an optimized algorithm is a good solution to enhance the FOV. In addition, with the development of computer, the combination of fast calculation algorithms and high-performance computing equipment is also an effective means to speed up the calculation and helpful in spatio-temporal multiplexing of SLMs. A path for limited FOV is opened up at the same time by the development of new optical devices. From the perspective of display for practical uses, more system parameters need to be taken into account, such as cost, image quality, uniform intensity distribution, and simplicity of overall system, among others. Holographic displays are anticipated to enter the market and become a part of daily life in the near future as a result of advancements in CGH algorithms, devices, and systems.

Author Contributions: M.R.: conceptualization, writing—original draft preparation methodology; N.J.: writing—original draft preparation; B.D.: supervision, methodology, writing—review and editing; R.K.: supervision, writing—review and editing, project administration, funding acquisition. All authors have read and agreed to the published version of the manuscript.

Funding: This research work is financially supported by the CSIR, New Delhi under the project MLP-2014.

Institutional Review Board Statement: Not applicable.

Informed Consent Statement: Not applicable.

Data Availability Statement: The data presented in this study are available on request from the corresponding author.

Conflicts of Interest: The authors declare no conflict of interest.

References

1. Geng, J. Three-dimensional display technologies. *Adv. Opt. Photon* **2013**, *5*, 456–535. [CrossRef] [PubMed]
2. Hariharan, P. *Optical Holography: Principles, Techniques and Applications*, 2nd ed.; Cambridge University Press & Assessment: Cambridge, UK, 2008; pp. 154–196.
3. Paturzo, M.; Memmolo, P.; Finizio, A.; Näsänen, R.; Naughton, T.J.; Ferraro, P. Holographic display of synthetic 3D dynamic scene. *3D Res.* **2010**, *2*, 31–35. [CrossRef]
4. Michalkiewicz, A.; Kujawinskaa, M.; Kozackia, T.; Wangb, X.; Bosb, P.J. Holographic three-dimensional displays with liquid crystal on silicon spatial light modulator. *Proc. SPIE* **2004**, *5531*, 85–94.
5. Pan, Y.; Liu, J.; Li, X.; Wang, Y. A review of dynamic holographic three-dimensional display: Algorithms, devices, and systems. *IEEE Trans. Ind. Inform.* **2015**, *12*, 1599–1610. [CrossRef]

6. Sando, Y.; Itoh, M.; Yatagai, T. Fast calculation method for cylindrical computer-generated holograms. *Opt. Express* **2005**, *13*, 1418–1423. [CrossRef] [PubMed]
7. Ma, Y.; Wang, J.; Wu, Y.; Jin, F.; Zhang, Z.; Zhou, Z.; Chen, N. Large field-of-view holographic display by gapless splicing of multisegment cylindrical holograms. *Appl. Opt.* **2021**, *60*, 7381–7390. [CrossRef] [PubMed]
8. Kozacki, T. Holographic display with tilted spatial light modulator. *Appl. Opt.* **2011**, *50*, 3579–3588. [CrossRef] [PubMed]
9. Liangcai, C.; Wang, Z.; Zhang, H.; Jin, G.; Gu, C. Volume holographic printing using unconventional angular multiplexing for three-dimensional display. *Appl. Opt.* **2016**, *55*, 6046–6051.
10. Liu, S.-J.; Xiao, D.; Li, X.W.; Wang, Q.H. Computer-generated hologram generation method to increase the field of view of the reconstructed image. *Appl. Opt.* **2018**, *57*, A86–A90. [CrossRef] [PubMed]
11. Wang, Z.; Tu, K.; Pang, Y.; Lv, G.Q.; Feng, Q.B.; Wang, A.T.; Ming, H. Enlarging the FOV of lensless holographic retinal projection display with two-step Fresnel diffraction. *Appl. Phys. Lett.* **2022**, *121*, 081103. [CrossRef]
12. Yaras, F.; Kovachev, M.; Ilieva, R.; Agour, M.; Onural, L. Holographic reconstructions using phase-only spatial light modulators. Presented at the 3DTV Conference: The True Vision—Capture, Transmission and Display of 3D Video, Istanbul, Turkey, 18 July 2008.
13. Maeno, K.; Fukaya, N.; Nishikawa, O.; Sato, K.; Honda, T. Electro-holographic display using 15mega pixels LCD. *Proc. SPIE* **1996**, *2652*, 15–23.
14. Yaraş, F.; Kang, H.; Onural, L. Multi-SLM holographic display system with planar configuration. Presented at the 2010 3DTV-Conference: The True Vision-Capture, Transmission and Display of 3D Video, Tampere, Finland, 7–9 June 2010.
15. Hahn, J.; Kim, H.; Lim, Y.; Park, G.; Lee, B. Wide viewing angle dynamic holographic stereogram with a curved array of spatial light modulators. *Opt. Exp.* **2008**, *16*, 12372–12386. [CrossRef] [PubMed]
16. Liu, S.-J.; Wang, D.; Zhai, F.X.; Liu, N.N.; Hao, Q.Y. Holographic display method with a large field of view based on a holographic functional screen. *Appl. Opt.* **2020**, *59*, 5983–5988. [CrossRef] [PubMed]
17. Duan, X.; Liu, J.; Shi, X.; Zhang, Z.; Xiao, J. Full-color see-through near-eye holographic display with 80 field of view and an expanded eye-box. *Opt. Exp.* **2020**, *28*, 31316–31329. [CrossRef] [PubMed]
18. Gu, Y.; Weng, Y.; Wei, R.; Shen, Z.; Wang, C.; Zhang, L.X.; Zhang, Y. Holographic waveguide display with large field of view and high light efficiency based on polarized volume holographic grating. *IEEE Photonics J.* **2021**, *14*, 1–7. [CrossRef]
19. Minseok, C.; Yoo, D.; Lee, S.; Lee, B. Etendue-Expanded Holographic Display Using Lens Array 4-f System. Presented at the Digital Holography and Three-Dimensional Imaging, Cambridge, UK, 1–4 August 2022.
20. Kim, D.; Nam, S.-W.; Lee, B.; Lee, B. Wide field of view holographic tiled display through axially overlapped holographic projection. Presented at the Ultra-High-Definition Imaging Systems V, San Francisco, CA, USA, 3 March 2022.

Proceeding Paper

Verification of SoC Using Advanced Verification Methodology [†]

Pranuti Pamula [1], Durga Prasad Gorthy [2], Phalguni Singh Ngangbam [1] and Aravindhan Alagarsamy [1,*]

[1] Multi-Core Architecture Computation (MAC) Lab, Department of ECE, Koneru Lakshmaiah Education Foundation, Vaddeswaram 522501, India; shellypranuti@gmail.com (P.P.); npsingh@kluniversity.in (P.S.N.)
[2] AMD, Hyderabad 500081, India; prasad.gorthy@gmail.com
*   Correspondence: drarvindhan@kluniversity.in
†   Presented at the International Conference on "Holography Meets Advanced Manufacturing", Online, 20–22 February 2023.

Abstract: The semiconductor industry has evolved significantly since its founding in 1950. Transistors and diodes are the primarily used electronic devices, but advancements in technology have led to more complex semiconductor devices, from printed circuit boards to multimillion gate design, i.e., a System on Chip (SoC) design. Almost 70–80 percent of the total SoC design effort is aimed at functional verification. In this paper, verification of an interconnect block in a processing system is presented. Trace monitoring of the transactions on the Advanced eXtensible Interface (AXI) interface of the interconnect is performed by programming different operational pointers and filters. Results were simulated from Synopsys—a Verilog Compiler Simulator (VCS) tool-2022v (Hyderabad, India).

Keywords: semiconductor industry; SoC; functional verification; AXI interface

Citation: Pamula, P.; Gorthy, D.P.,
Ngangbam, P.S.; Alagarsamy, A.
Verification of SoC Using Advanced
Verification Methodology. *Eng. Proc.*
2023, *34*, 12. https://doi.org/
10.3390/HMAM2-14160

Academic Editor: Vijayakumar
Anand

Published: 13 March 2023

1. Introduction

In recent years, the complexity of System on Chip (SoC) design has increased. The higher number of components in a single chip makes the verification of any SoC design very critical. Hence, a proper methodology for any SoC or IP is required [1–4]. Despite all the advancements, there is a significant gap between the modern technology and the verification needs of new industries. This situation is becoming worse regarding to the change in design as there is rapid movement towards the era of automated vehicles, smart cities and the Internet of Things (IoT) [5–8]. Moreover, these electronic devices collect personal information such as location, sleep patterns, health, etc., which is stored in the billions of computer devices that operate without pause and even the surrounding environment may have compromised or malicious devices. As the system design and security have transformed to adapt themselves, so must the verification adjust as well. Regarding the growing requirements for the design and the time to market, the duration has shrunk from years required for verification and hard work to less than a year. This aggressive shrinking implies shorter timespan for a thorough design review, which may cause misunderstanding of the requirements and a consequent increase in errors. Therefore, verification is expected to handle more errors in design with even less time duration.

Problem Statement

Verification of an SoC design is carried out at different stages with a different approach as per design specifications. With interconnect as a common ground for all the rest of the design to interact, there are many functionalities to be verified and connectivity checks to be conducted. Many tests need to be developed for the verification of an interconnect. Connectivity checks at the interface interconnect being the most important requires detailed analysis and thorough research of the design specification. Track sourcing from the primary interface should reach the desired secondary interface without any loss in the data packets, such checks between multiple primary and secondary interfaces are carried out by initiating write operations to a register space of the secondary interface and expecting to read the same data without any errors. These transactions can be self-tested using System Verilog assertions and checkers. The other functionalities of the design can be verified with a variety of approaches.

2. Materials and Methods

Deep sub-micron effects complicate design closure for very large designs [9]. A System on Chip (SoC) is an IC (Integrated Circuit) which is designed by integrating multiple standalone VLSI designs that provide complete functionality for an application. SoC integrates a microprocessor with advanced peripherals such as a coprocessor, memory elements, GPU, Wi-Fi module, etc. This definition of SoC emphasizes the predesigned models of complex design functions which are known as cores. These can be intellectual property blocks, virtual components, macros, etc. In SoC, in-house library cores may be used along with some cores designed by other design houses known as intellectual property. Because of the use of the embedded software and the increasing integration of cores, the design complexity of SoC has increased dramatically over the past few years. In addition, it is still expected to grow at a very fast rate. According to Moore's law, silicon complexity quadruples every three years [10]. This complexity accounts for the huge size of cores and shrinking geometry.

There are three types of SoCs that are totally distinguishable, i.e., a SoC built around a micro-controller, a SoC built around a microprocessor, and a programmable SoC, where the internal elements are not predefined and can be programmed in any essential manner. These kinds of SoCs are also known as FPGAs or complex programmable logic devices. In all SoC designs, predefined cores are the essential components. The flexibility of the cores depends on the form in which they are available. The trade-off between these cores is in terms of performance, power, speed, area, flexibility, cost, time to market, etc. [11].

2.1. Architectural Overview

The SoC architecture integrates a feature of a dual- or single-core microprocessor core-based processing system and a Xilinx programmable logic in a single device. It is built on state-of-the-art technology that offers high performance and low power [12]. The multi-core processors are the heart of the PS, which also includes on-chip memory, external memory interfaces, and a rich set of I/O peripherals.

SoC offers the flexibility and scalability of an FPG, while providing performance, power, and ease of the use. The range of devices in the family of SoC enables the designers to find cost-sensitive as well as high-performance applications from a single platform using standard tools.

Functional blocks of SoC are shown in Figure 1. The processing system and programmable logic both operate on different power domains. This configuration enables the users to manage the power utilization of PL if required.

The SoC is composed of two major functional blocks:

- Processing system;
- Programmable logic.

Processing System (PS)

- Application processor unit: The application processor unit offers high performance and standard-compliant capabilities. The runtime configurations allow the single processor or asymmetrical or symmetrical multiprocessing setups. It is a 32 Kb instruction set with a 32 Kb cache [12]. In addition, a sharable 512 Kb cache with parity is available. An accelerator coherency port from PL to PS, which is a 64 b AXI slave port, provides a connection between the processing system and programmable logic. APU also contains 256 Kb of on-chip SRAM which is a dual-ported memory. It is accessible to CPUs, PL, and central interconnects. There are four DMA (direct memory access) channels for PS to copy data from CPU memory to/from other system memories.

- Memory interfaces: The memory interface of PS includes multiple memory technologies. It consists of DDR controllers with 16 b and 32 b widths. This uses up to 73 dedicated pins of PS [12]. The DDR can be powered down as per the idle periods of PS. Transaction scheduling is performed for optimizing data bandwidth and latency. The efficiency of memory is increased by 90% by advanced re-ordering engines and increased by 80% with random read/write operations. Collision check monitors the memory for any write–read collisions and the write buffer is used in that case. The primary boot device can be a NAND controller or a parallel SRAM/NOR controller.
- I/O peripherals: The input–output peripherals are a collection of industry standard interfaces for communication with external systems. Programmable interrupt controllers on the GPIO are used for a status read of raw and masked interrupts. These interrupts are positive-edge, negative-edge, either–edge, high-level, or low-level sensitive. A USB 2.0 high-speed on-the-go (OTG) dual-role USB host controller and USB device controller operations are performed using a single hardware. This configuration uses MIO pins only. The USB host controller registers, and data structures are EHCI compatible [12]. It supports up to 12 endpoints.
- Interconnect: The SoC uses several interconnect technologies that are optimized for specific communication requirements of the functional blocks. The SoC interconnect is divided into two parts: one is based on the high-performance data path of AXI on the PS interconnect and the other is based on PS-PL interfaces. The PS interconnect consists of an OCM interconnect and a central interconnect. The OCM interconnect provides access to 256 Kb of memory from the central interconnect and PL. The CPU and ACP interfaces have the lowest latencies to OCM through the SCU. The central interconnect is a 64-bit interconnect. It connects the input–output peripherals and the DMA controller to the DDR memory controller, on-chip RAM, and the AXI-GP interfaces for PL logic. It also connects the local DMA units in Ethernet, USB, and SD/SDIO controllers to the central interconnect. It also connects the PS master to the IOP.

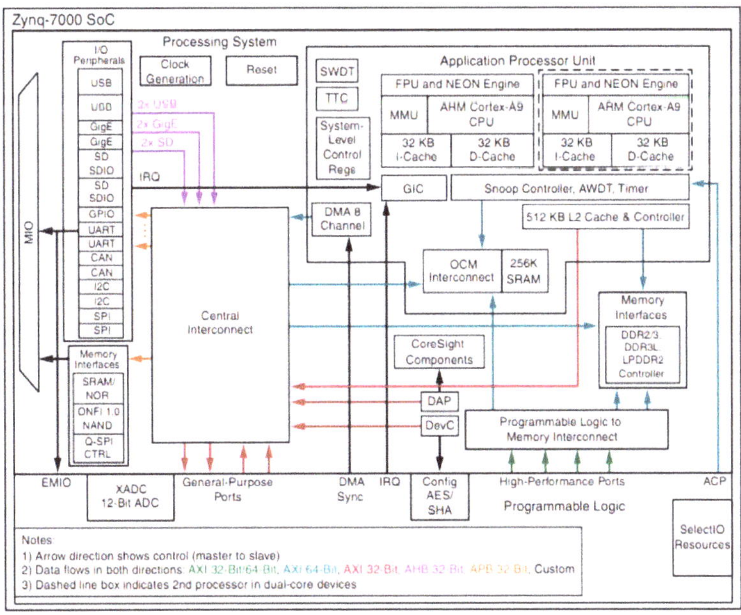

Figure 1. SoC architecture.

2.2. Connectivity Check in Interconnect

Interconnect consists of several input and output interfaces. Each of the interfaces reaches out to different slave modules or input from connected masters. The connectivity checks are essential at every interface. This is performed to verify the transactions that are intended to pass through a particular interface are reaching without any loss of data packets. Such checks are carried out by initiating from a master to register space of slave and expect to read exactly the same data without any errors. This behavior of the design is verified by using the system Verilog Assertions and data comparison using C or System Verilog [13]. The analysis of the simulated result is as important as defining the sequence of the test. The occurrence of an error or unexpected behavior at the output is required to be traced back to the source of the issue. A large amount of time is spent on debugging of simulated results. One of the test scenarios generated to verify connectivity at an interface is as discussed. The simulation result for verification of an AXI interface and APB interface are shown in the waveforms.

3. Results

Verification of blocks is of utmost importance. This is achieved by programming a testbench to monitor outgoing traffic with the help of pointers. This outgoing traffic can contain a large bandwidth of data signals. These outgoing data can be filtered as per the requirement and a trace can be generated for only those selected data signals or transactions. The pointers required to monitor the interface are programmed using a set of control registers. These configurations are shown in Figure 2.

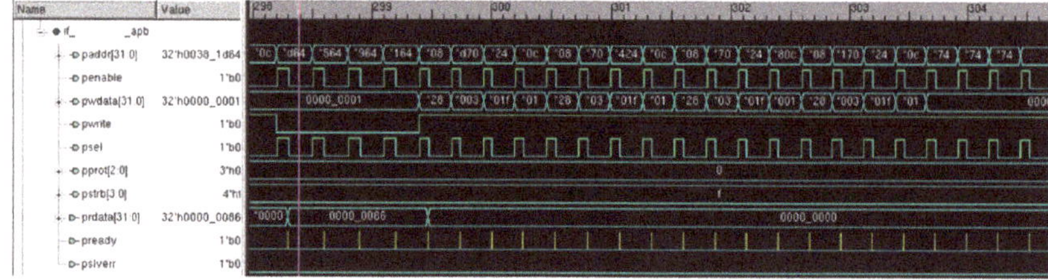

Figure 2. Configuration of pointers.

When pointers are configured and set to monitor a port, the filters are activated to filter out the required amount of data. The filters are configured as per the address or ID of the transactions and later subdivided by control/instruction trace, data trace, bus trace, interface trace, fabric trace, etc. Then, a burst of AXI transactions is sent to the observed port. A set of such transactions is displayed in Figures 3–5.

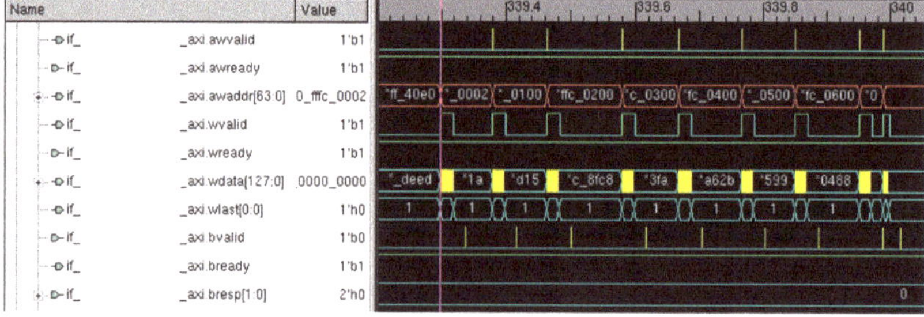

Figure 3. Write operation of the AXI burst transfer.

Figure 4. Read operation of the AXI burst transfer.

Figure 5. Read response on the AXI burst transfer.

The write and read burst transfers sent to the AXI bus are shown in Figure 6. The above image shows traffic sent to the observed port. These transactions are then observed with the help of pointers. The transactions on the interface are filtered and sent out to the counter, to count the number of transaction hits. The output is thus observed and analyzed for verification.

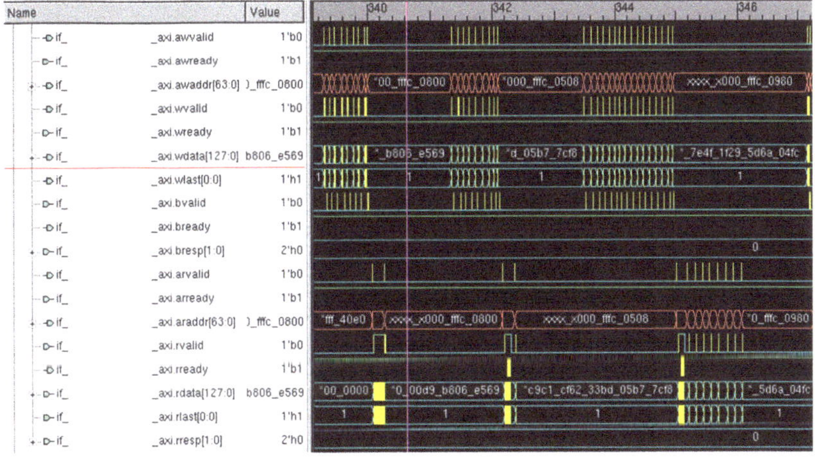

Figure 6. Burst traffic on the AXI interface.

Below are the waveforms depicting the counter values at the output. The output of the write request pointer is shown in Figure 7.

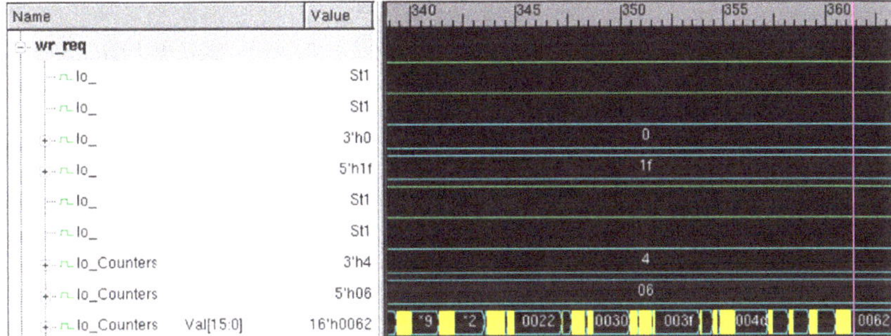

Figure 7. Write request transfers.

The output of the write response pointer is shown in Figure 8.

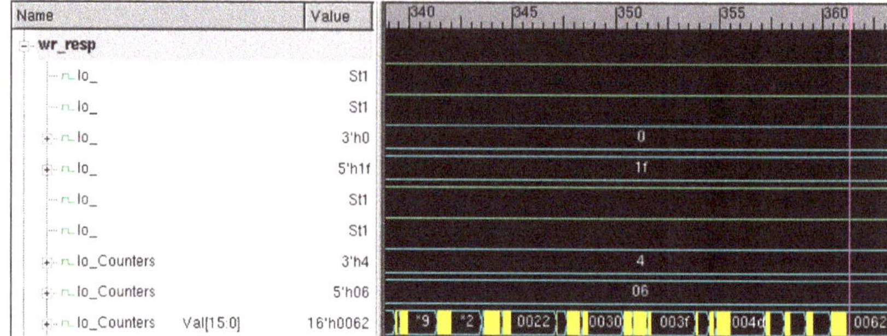

Figure 8. Write response transfers.

The output of the read request pointer is shown in Figure 9.

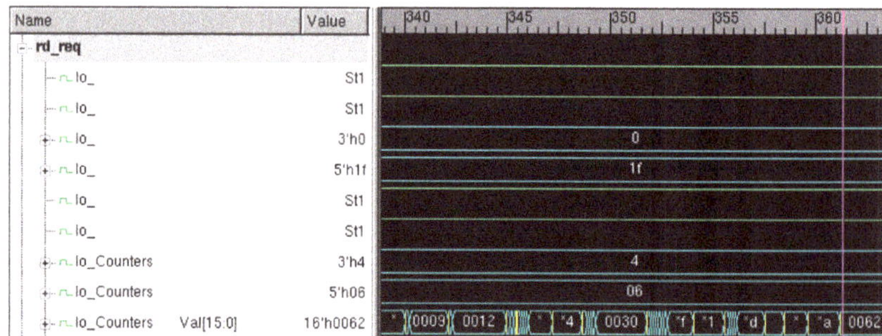

Figure 9. Read request transfers.

The output of the read response pointer is shown in Figure 10.

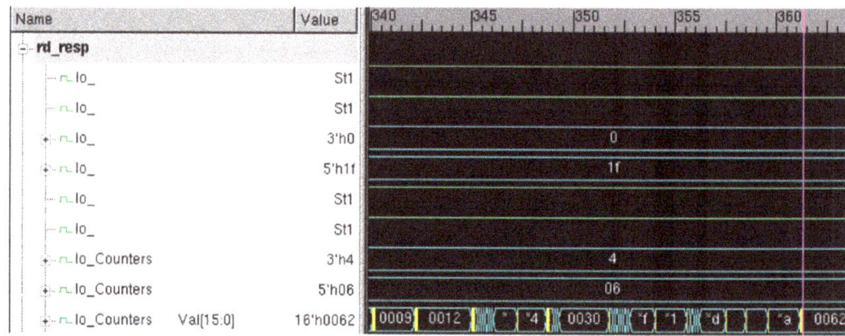

Figure 10. Read response transfers.

- Green line: Transaction of 1.
- Blue line: Transaction of 0.
- Yellow line: Brust Transactions.
- Purple line: Marker.
- Red line: transaction of address is represented.

4. Conclusions

SoC verification is a complex and never-ending task. The process can be faster and more efficient when proper programming and simulation tools are used. Verification can be achieved with prior knowledge of SoC architecture and RTL design, where the environment is built using UVM and System Verilog. All the parts of the testbench can be reused easily for different designs. This reduces verification complexity and improves efficiency. The design functionalities are verified by using assertions and checkers along with the basic test sequence.

The connectivity of an interconnect block with several interfaces is verified successfully. The performance monitoring at various interfaces of interconnect is successfully completed. The simulation results are compared and evaluated by a self-checking testbench. This reduces extra efforts to locate the problem or issue in the design, as it locates the exact timestamp and points at the exact line of the RTL code where a violation has occurred.

Author Contributions: A.A.: Conceptualization, methodology, software, investigation, writing—original draft, funding acquisition, project administration; P.P.: Methodology, formal analysis, writing—original draft; D.P.G.: Investigation, software, resources; P.S.N.: Supervision. All authors have read and agreed to the published version of the manuscript.

Funding: A.A.: P.P.: P.S.N.: The first authors thanks DST-FIST for funding the lab facility for supporting this research under grant number SR/FST/ET-II/2019/450.

Institutional Review Board Statement: Not applicable.

Informed Consent Statement: Not applicable.

Data Availability Statement: No data was used for the research described in the article.

Conflicts of Interest: The authors declare that they have no known competing financial interest or personal relationships that could have appeared to influence the work reported in this paper.

References

1. Ghosh, P.; Srivastava, R. Case Study: SoC Performance Verification and Static Verification of RTL Parameters. In Proceedings of the 2019 20th International Workshop on Microprocessor/SoC Test, Security and Verification (MTV), Austin, TX, USA, 9--10 December 2019; pp. 65–72.
2. Huang, X.; Liu, L.; Li, Y.; Liu, L.; Huang, X. FPGA Verification Methodology for SiSoC Based SoC Design. In Proceedings of the 2011 IEEE International Conference of Electron Devices and Solid-State Circuits, Tianjin, China, 17–18 November 2011.
3. Bai, L.; Fan, X.; Zhang, M.; Sun, L. A VMM/FPGA Co-verification Method for "Longtium Stream" Processor. In Proceedings of the 2013 IEEE International Conference on Signal Processing, Communication and Computing (ICSPCC 2013), KunMing, China, 5–8 August 2013.
4. Podivinsky, J.; Simkova, M.; Cekan, O.; Kotasek, Z. FPGA Prototyping and Accelerated Verification of ASIPs. In Proceedings of the 2015 IEEE 18th International Symposium on Design and Diagnostics of Electronic Circuits & Systems, Belgrade, Serbia, 22–24 April 2015; pp. 145–148.
5. Noami, A.; Alahdal, A.; Kumar, B.P.; Chandrasekhar, P.; Safi, N. High Speed Data Transactions for Memory Controller Based on AXI4 Interface Protocol SoC. In Proceedings of the 2021 International Conference on Advances in Electrical, Computing, Communication and Sustainable Technologies (ICAECT), Bhilai, India, 19–20 February 2021.
6. Seongyoung, S.; Moon, J.; Jun, S. FPGA-Accelerated Time Series Mining on Low-Power IoT Devices. In Proceedings of the 2020 IEEE 31st International Conference on Application-specific Systems, Architectures and Processors (ASAP), Manchester, UK, 6–8 July 2020.
7. Noami, A.; Kumar, B.P.; Paidimarry, C.S. Power Optimization for Multi-Core Memory Controller Using Intelligent Clock Gating Technique. *J. Electr. Electron. Eng.* **2022**, *15*, 129–137.
8. Gophane, K.C.; Bhaskar, P.C. FPGA Based Adaptive IoT Framework for Distinct Applications. In Proceedings of the 2018 Fourth International Conference on Computing Communication Control and Automation (ICCUBEA), Pune, India, 16–18 August 2018.
9. Accounting For Very Deep Sub-Micron Effects in Silicon Models. Available online: https://www.eetimes.com/accounting-for-very-deep-sub-micron-effects-in-silicon-models/ (accessed on 15 February 2023).
10. Tuomi, I. The lives and death of Moore's Law. *First Monday* **2002**, *7*. [CrossRef]
11. Noguera, J.; Badia, R.M. System-level power-performance trade-offs in task scheduling for dynamically reconfigurable architectures. In Proceedings of the 2003 International Conference on Compilers, Architecture and Synthesis for Embedded Systems, San Jose, CA, USA, 30 October–1 November 2003.

12. Zynq-7000 SoC Data Sheet: Overview. Available online: https://docs.xilinx.com/v/u/en-US/ds190-Zynq-7000-Overview (accessed on 15 February 2023).
13. Chris, S. *System Verilog for Verification: A Guide to Learning the Testbench Language Features*; Springer Science & Business Media: Berlin/Heidelberg, Germany, 2008.

Proceeding Paper

Techniques to Expand the Exit Pupil of Maxwellian Display: A Review [†]

Kaur Rajveer [1,2] and Kumar Raj [1,2,*]

[1] CSIR-Central Scientific Instruments Organisation, Sector 30C, Chandigarh 160030, India; 17rajveer.kaur@gmail.com
[2] Academy of Scientific and Innovative Research (AcSIR), Ghaziabad 201002, India
[*] Correspondence: raj.optics@csio.res.in
[†] Presented at the International Conference on "Holography Meets Advanced Manufacturing", Online, 20–22 February 2023.

Abstract: Near-eye display (NED) devices are required to provide visual instructions in the fields of education, navigation, military operations, construction, healthcare, etc. The issues with conventional NEDs are the form factor and vergence–accommodation conflict (VAC). The Maxwellian display alleviates the VAC in NEDs by providing consistently focused virtual images to the viewer, regardless of the depth of focus of the human eye. The main limitation of the Maxwellian display is its limited exit pupil size. Due to misalignment of the device or eyeball rotation, the user may miss the eye box, and the image will become lost. To mitigate this limitation, exit pupil expansion can be obtained either statically or dynamically. This paper reviews the various techniques employed to expand the exit pupil. The review includes the principle, advantages, and drawbacks of various techniques for expanding the exit pupil of the Maxwellian display. The structure of the paper starts with an introduction and the principle of the Maxwellian display, followed by a discussion of the main limitations that arise with various techniques, along with potential solutions.

Keywords: Maxwellian display; exit pupil; near-eye display; AR/VR display

1. Introduction

The development of augmented reality (AR) technology has gained a lot of attention in recent years or decades. Augmented reality, which combines digital images with a physical world in three dimensions (3D), has recently grown in popularity in the scientific field [1]. The near-eye display is the most popular AR device due to its best immersive effect and portable design. NED devices are employed in the fields such as architecture, construction, education, navigation, military operations, and gaming [1]. Since existing NEDs focus virtual images on a fixed focus plane, there is a fixed accommodation distance. The distance from the focus plane to the NEDs does not match the vergence distance, known as vergence–accommodation conflict (VAC), which results in discomfort and visual fatigue [2]. The VAC issue is resolved by the Maxwellian display, offering consistently in-focus images to the user independent of the optical power of the human eye [3]. The Maxwellian display is based on the Maxwellian view, in which projection of images directly onto the retina is carried out by focusing the light rays on the eye's pupil instead of providing the proper depth cues. The light beams from the display are converged by the eyepiece lens at the eye pupil plane and then projected onto the retina. However, the eye box size is limited to the size of the eye pupil [2,3]. This tiny eye box is uncomfortable for the wearer because a slight misalignment of the device or eyeball rotation makes the image disappear entirely. However, the limited eye box of Maxwellian displays remains a major challenge in the development of AR. Several methods have been proposed to enlarge the eye box.

This review includes various techniques to expand the exit pupil in active and passive ways. The review includes the principle, advantages, and drawbacks of various approaches

Citation: Rajveer, K.; Raj, K. Techniques to Expand the Exit Pupil of Maxwellian Display: A Review. *Eng. Proc.* **2023**, *34*, 13. https://doi.org/10.3390/HMAM2-14128

Academic Editor: Vijayakumar Anand

Published: 6 March 2023

for expanding the exit pupil of the Maxwellian display that have been reported in the literature, and is divided into three sections: static displays, tunable displays, and dynamic displays.

2. Techniques

2.1. Static Viewpoints

Exit pupil expansion of Maxwellian displays can be achieved by generating multiple viewpoints. Multiple viewpoints can be generated using the multiplexing technique. As an enlarged eye box is crucial for comfortable viewing in NEDs, angular multiplexing is used in Maxwellian displays. Multiple concave mirrors are recorded in a holographic optical element (HOE) as an out-coupler of a waveguide [4]. The HOE focuses the displayed images into multiple spots in the eye pupil plane, enhancing the exit pupil in the configuration. However, in this work, the exit pupil expansion is only carried out horizontally. The exit pupil can be extended in a vertical direction using the vertical high diffraction orders of the spatial light modulator (SLM) by enlarging the aperture in the Fourier plane of the 4-f system [5]. In the study reported by Shrestha et al., an array of beam splitters was used to expand the exit pupil, which resulted in an increase in weight or form factor due to bulky optics [6]. In the previous research, setups are bulky and have a high cost. A computational imaging-based holographic Maxwellian near-eye display addressed these shortcomings [7]. In this paper, encoding a complex wavefront into amplitude-only signals produces an all-in-focus virtual image. The hologram is multiplexed with several off-axis plane waves, which duplicate the pupils into an array to enlarge the exit pupil. As a result, this approach has a small form factor and only needs one active electrical component, which supports wearable applications. The aforementioned methods are difficult to use in a full-color NED due to the narrow bandwidth of diffractive optical elements [8]. A Maxwellian NED in full color with an expanded exit pupil in two-dimensional (2D) space is used to overcome the tiny eye box restriction [2]. With the use of a quarter-wave plate (QWP) and two Pancharatnam–Berry deflectors (PBDs), a broadband 2D beam deflector can be used to multiplex the one viewpoint into a 3×3 array of viewpoints, as shown in Figure 1a. There is a demerit of aforementioned methods in that each viewpoint image is overlapped, or the area becomes blank when changing the position from one viewpoint to another viewpoint. In order to prevent the images from overlapping on the retina, this effect is removed by creating multiple independent viewpoints [9]. In this method, a high-speed MEMS mirror can be used as an aperture stop for a narrow exit pupil, which provides many views based on the time-multiplexing approach, and gives a continuous image over a wide eye box without an eye tracker. To acquire a different area and perspective of a scene to each viewpoint, multiple HOEs are designed accordingly and spatially located. These viewpoint images do not overlap on the retina.

2.2. Tunable Viewpoints

Most Maxwellian NEDs with eye box replication give fixed intervals between the viewpoints, which must be tunable according the variation in eye pupil size among users. Focal spot steering is accomplished by synthesizing the CGH with various plane carrier waves [10]. In this method, the transverse position offset in the plane of the eye's pupil depends on the spatial frequency of the plane wave. By changing the frequency of carrier wave in the CGH synthesis, multiple focal spots can be steered, as shown in Figure 1b. There is another technique to extend the eye box in Maxwellian see-through NEDs, which uses polarized gratings (PGs) and a multiplexed HOE [11]. The transmission PGs selectively diffract light beams with different polarization states and have high diffraction efficiency in ±1 orders. Two viewpoints are generated by the multiplexed HOE and are further duplicated to four viewpoints at different locations by using these two PGs, as shown in Figure 1c. By mechanically moving the PG, these viewpoints can be tuned accordingly. However, the tuning of image viewpoints and image shifts is still limited, which hinders its practicability. To overcome this limitation, an adjustable and continuous replication of eye box of a holographic Maxwellian near-eye display system has been proposed, in which different frequencies are employed to the hologram using spatial multiplexing to guide the beams in the required directions [12]. This allows us to create a pupil array, generate the sub-holograms corresponding to the viewpoints, then multiplex them all to make the composite hologram. The interval of the focus spots is dynamically adjusted, making it feasible to adapt to any eye pupil size and prevent the image overlapping and blind region issues. To minimize the operating speed and the bulky form factor, and to eliminate the problem with double or blank images, Yoo et al. [13] designed a light guide based on switchable viewpoints in the Maxwellian display. With this method, using the polarization grating, multiplexed HOEs, and polarization-dependent eyepiece lens, the expansion of the eye box is carried out without an additional mechanical movement of elements, as represented in Figure 1d. With the polarization-multiplexing approach, the polarizer rotator independently activates two different groups of viewpoints and is synchronized with the eye tracker.

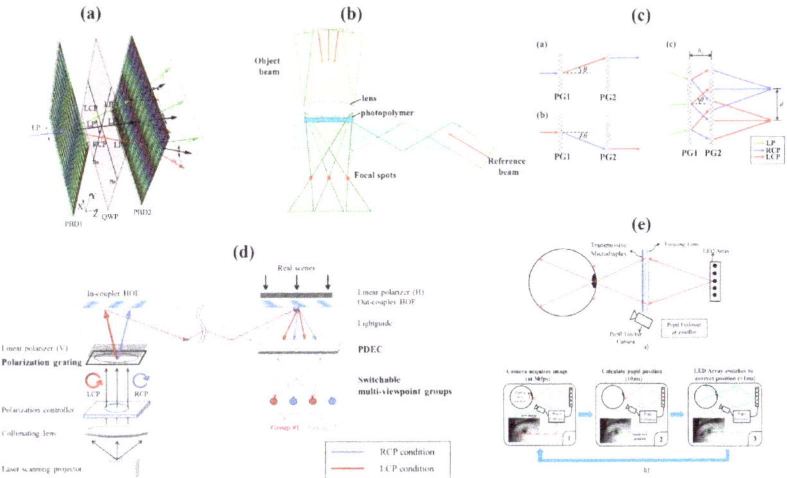

Figure 1. (**a**) Eye box expansion using polarization-dependent HOEs (Reprinted with permission from Ref. [2]. Copyright 2020 The Optical Society); (**b**) Eye box expansion using multiplexed HOE (Reprinted with permission from Ref. [10]. Copyright 2018 The Optical Society); (**c**) Tunable viewpoints using polarization-dependent HOEs (Reprinted with permission from Ref. [11]. Copyright 2021 The Optical Society); (**d**) Switchable eye box using polarization multiplexing (Reprinted with permission from Ref. [13]. Copyright 2020 The Optical Society); (**e**) Viewing point steering with backlight modulation using LED array (Reprinted with permission from Ref. [14]. Copyright 2019 The Optical Society).

2.3. Dynamic Viewpoints

Most of the Maxwellian display setups in the previous section do not include an eye-tracking device, which is needed for more realistic applications. By incorporating pupil-tracking, exit-pupil steering has been proposed as a light field projection-type display [15]. However, exit-pupil shifting for a holographic display has not yet been developed. Takaki et al. proposed a technique based on wave optics [16]. This technique can electrically modify the light's convergence point in response to an eye's movement without needing mechanical components, and produces a steerable eye box. An eye pupil-tracking Maxwellian system using a steering mirror and a pupil-shifting HOE is proposed, which provides a dynamic eye box that can be accomplished by using the pupil-shifting HOE combiner to laterally shift a view point in order to follow the movement of eye pupil [17]. However, the eye box is shifted accordingly by changing the incidence angle to the HOE in accordance with the detected eye location. Although the eye box is capable of 2D shifting, the possible incidence angle range is constrained by the aberrations and angular selectivity, which also restricts the eye box's shifting range. In the study reported by Kim et al. [18], the dynamic eye box is generated by laterally translating the HOE. Both of the aforementioned techniques make use of the optical element's mechanical movement, which results in an increase in the weight or form factor. The backlight modulation technique has been reported in the literature to generate a dynamic eye box in full-color Maxwellian displays [14]. An array of light-emitting diodes (LEDs) and a pupil tracker are synchronized to generate a steerable eye box based on the pinhole imaging principle represented in Figure 1e. This approach can switch fixed focal spots with low motion-to-photon latency. The aforementioned pupil-steering techniques rely on modifying the incident light angle. However, the lens coupler's diffraction-limited performance is only achievable at one incidence angle; these techniques result in significant aberration. A new pupil-steering approach employs a switchable polarization converter and a cholesteric liquid crystal holographic lens (CLCHL) [19]. The polarization converter controls the light's polarization and chooses the appropriate holographic lens to operate, as opposed to depending on changing the incidence angle to change the focal point.

3. Conclusions

This paper reviewed the state-of-the-art Maxwellian display design, focusing on two comfort features, form factor and a large eye box. We have introduced the conventional Maxwellian display and its principles, and then discussed the multiplexing techniques available to expand the exit pupil. Techniques for enlarging the exit pupil of Maxwellian displays, such as spatial and angular multiplexing of HOEs, polarization multiplexing, backlight modulation, and materials, are reviewed. Our paper discusses the relative merits and demerits of the methods, along with potential solutions in terms of achieving the goals of AR displays.

Author Contributions: K.R. (Kaur Rajveer): writing—original draft preparation, methodology; K.R. (Kumar Raj): supervision, funding acquisition, writing—review and editing, project administration. All authors have read and agreed to the published version of the manuscript.

Funding: This research was funded by JRF, CSIR, India, and project number MLP2014, CSIR, India.

Institutional Review Board Statement: Not applicable.

Informed Consent Statement: Not applicable.

Data Availability Statement: The data presented in this study are available on request from the corresponding author.

Conflicts of Interest: The authors declare no conflict of interest.

References

1. He, Z.; Sui, X.; Jin, G.; Cao, L. Progress in virtual reality and augmented reality based on holographic display. *Appl. Opt.* **2019**, *58*, A74–A81. [CrossRef] [PubMed]
2. Lin, T.; Zhan, T.; Zou, J.; Fan, F.; Wu, S. Maxwellian near-eye display with an expanded eyebox. *Opt. Express* **2020**, *28*, 38616–38625. [CrossRef] [PubMed]
3. Westheimer, G. The Maxwellian view. *Vision Res.* **1966**, *6*, 669–682. [CrossRef] [PubMed]
4. Kim, S.B.; Park, J.H. Optical see-through Maxwellian near-to-eye display with an enlarged eyebox. *Opt. Lett.* **2018**, *43*, 767–770. [CrossRef] [PubMed]
5. Choi, M.H.; Ju, Y.G.; Park, J.H. Holographic near-eye display with continuously expanded eyebox using two-dimensional replication and angular spectrum wrapping. *Opt. Express* **2020**, *28*, 533–547. [CrossRef] [PubMed]
6. Shrestha, P.K.; Pryn, M.J.; Jia, J.; Chen, J.S.; Fructuoso, H.N.; Boev, A.; Zhang, Q.; Chu, D. Accommodation-free head mounted display with comfortable 3D perception and an enlarged eye-box. *Research* **2019**, *2019*, 9273723. [CrossRef] [PubMed]
7. Chang, C.; Cui, W.; Park, J.; Gao, L. Computational holographic Maxwellian near-eye display with an expanded eyebox. *Sci. Rep.* **2019**, *9*, 18749. [CrossRef] [PubMed]
8. Kaur, R.; Pensia, L.; Singh, O.; Das, B.; Kumar, R. Study of wavelength dependency on diffraction efficiency of volume holographic gratings based couplers in waveguide displays. In *Computational Optical Sensing and Imaging*; Optica Publishing Group: Vancouver, BC, Canada, 2022.

9.	Jo, Y.; Yoo, C.; Bang, K.; Lee, B.; Lee, B. Eye-box extended retinal projection type near-eye display with multiple independent viewpoints. *Appl. Opt.* **2021**, *60*, A268–A276. [CrossRef] [PubMed]
10.	Park, J.H.; Kim, S.B. Optical see-through holographic near-eye-display with eyebox steering and depth of field control. *Opt. Express* **2018**, *26*, 27076–27088. [CrossRef] [PubMed]
11.	Shi, X.; Liu, J.; Zhang, Z.; Zhao, Z.; Zhang, S. Extending eyebox with tunable viewpoints for see-through near-eye display. *Opt. Express* **2021**, *29*, 11613–11626. [CrossRef] [PubMed]
12.	Zhang, S.; Zhang, Z.; Liu, J. Adjustable and continuous eyebox replication for a holographic Maxwellian near-eye display. *Opt. Lett.* **2022**, *47*, 445–448. [CrossRef] [PubMed]
13.	Yoo, C.; Chae, M.; Moon, S.; Lee, B. Retinal projection type lightguide-based near-eye display with switchable viewpoints. *Opt. Express* **2020**, *28*, 3116–3135. [CrossRef] [PubMed]
14.	Hedili, M.K.; Soner, B.; Ulusoy, E.; Urey, H. Light-efficient augmented reality display with steerable eyebox. *Opt. Express* **2019**, *27*, 12572–12581. [CrossRef] [PubMed]
15.	Jang, C.; Bang, K.; Moon, S.; Kim, J.; Lee, S.; Lee, B. Retinal 3D: Augmented reality near-eye display via pupil-tracked light field projection on retina. *ACM Trans. Graph.* **2017**, *36*, 1–13. [CrossRef]
16.	Takaki, Y.; Fujimoto, N. Flexible retinal image formation by holographic Maxwellian-view display. *Opt. Express* **2018**, *26*, 22985–22999. [CrossRef] [PubMed]
17.	Jang, C.; Bang, K.; Li, G.; Lee, B. Holographic near-eye display with expanded eye-box. *ACM Trans. Graph.* **2018**, *37*, 1–14. [CrossRef]
18.	Kim, J.; Jeong, Y.; Stengel, M.; Aksit, K.; Albert, R.A.; Boudaoud, B.; Greer, T.; Kim, J.; Lopes, W.; Majercik, Z. Foveated AR: Dynamically-foveated augmented reality display. *ACM Trans. Graph.* **2019**, *38*, 99. [CrossRef]
19.	Xiong, J.; Li, Y.; Li, K.; Wu, S.T. Aberration-free pupil steerable Maxwellian display for augmented reality with cholesteric liquid crystal holographic lenses. *Opt. Lett.* **2021**, *46*, 1760–1763. [CrossRef] [PubMed]

Proceeding Paper

Design of Efficient Phase Locked Loop for Low Power Applications [†]

Chandra Keerthi Pothina *, Ngangbam Phalguni Singh *, Jagupilla Lakshmi Prasanna, Chella Santhosh * and Mokkapati Ravi Kumar

Department of Electronics and Communication Engineering, Koneru Lakshmaiah Education Foundation, Guntur 522302, India; lakshmiprasannanewmail@kluniversity.in (J.L.P.); ravikumar@kluniversity.in (M.R.K.)
* Correspondence: ckeerthip7@gmail.com (C.K.P.); npsingh@kluniversity.in (N.P.S.); csanthosh@kluniversity.in (C.S.)
† Presented at the International Conference on "Holography Meets Advanced Manufacturing", Online, 20–22 February 2023.

Abstract: The phase-locked loop is a technique that has contributed significantly to technological advancements in many applications in the fast-evolving digital era. In this paper, a Phase Locked Loop (PLL) is designed using 90 nm CMOS technology node with 1.8 V supply voltage. It features a PLL design with minimum power consumption of 194.26 µW with better transient analysis and DC analysis in an analog-to-digital environment. The proposed PLL design provides the best solution for many applications where a PLL is required with high performance but has to be accommodated in less area and low power consumption than state-of-the-art methods. This PLL not only works at high speed but also makes whole system work at low power in a very effective manner, which suits the present digital electronics circuits.

Keywords: Phase Locked Loop (PLL); Phase Frequency Detector/Charge Pump (PFD/CP); Low Pass Filter (LPF); Current Starved Voltage-Controlled Oscillator (CSVCO); Analog Digital Environment (ADE)

1. Introduction

The current era is all about compact battery devices used in electronic devices. A PLL is used in all SOC devices where circuitry generates a system clock signal. Basically, a PLL has a feedback loop that controls the phase of the output signal with the input signal along with phase error.

Figure 1 shows the block diagram of a PLL. It mainly consists of four blocks namely Phase Frequency Detector (PFD), Charge Pump with Low Pass Filter, a Voltage-Controlled Oscillator (VCO) to provide oscillations and a Frequency Divider [1].

Citation: Pothina, C.K.; Singh, N.P.; Prasanna, J.L.; Santhosh, C.; Kumar, M.R. Design of Efficient Phase Locked Loop for Low Power Applications. *Eng. Proc.* **2023**, *34*, 14. https://doi.org/10.3390/HMAM2-14157

Academic Editor: Aravind Rajeswary

Published: 13 March 2023

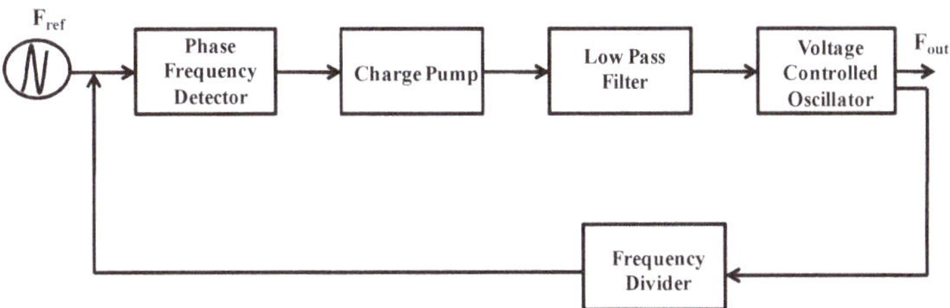

Figure 1. Block diagram of PLL.

2. Methodology

Previously, a multiplier was used as an analog phase detector, but it had a limited blocking range with a phase error of more than $90°$ where the output voltage is reduced. A digital phase detector was implemented where the mean value is proportional to the phase error [2]. The X-OR gate was used as a phase detector, which is linear, but if it were in phase and the error was greater than $180°$, it would lose its linearity. Consequently, the phase-frequency detectors are designed to detect phase and frequency differences, which increases the speed of PLL. For the LPF design, first order filter could be used, but it would introduce ripples as the control voltage jumps high when current is injected from the charge pump. To solve this problem, a second-order low-pass filter was designed to suppress the generated ripple. Firstly, VCOs (Voltage Controlled Oscillators) were designed using LC oscillators or ring oscillators [3]. However, ring oscillators are not stable as they let the switching characteristics of logic gates fluctuate by $+/-20\%$ and the disadvantage of LC oscillators has more matrix area. Therefore, the current starved VCOs are realized in our proposed design.

The PLL works in three states: Free Running state, Capture state, and Phase Lock state. As the name suggests, the free-running state refers to the phase in which no input voltage is present. As soon as the input frequency is applied, the VCO starts switching and starts producing an output frequency to be compared, and this stage is called the capture state [4].

The frequency comparison stops as soon as the output frequency equals the input frequency. This phase is known as the phase-locked state. PLL is a convenient circuit block widely used in wireless applications and electronics, from cell phones to radios, televisions, Wi-Fi routers, areas like FM demodulators, AM demodulators, frequency synthesizers, and more recovery, etc. [5].

3. Design and Implementation

The proposed design of 90 nm PLL was composed using cadence virtuoso and consisted of all the blocks of a Basic PLL. In this design, the number of transistors is reduced when compared to the existing designs in [6]. Due to this, the area is reduced. The reduced transistors also lead to a decreased usage of the power consumption and cost. The design of blocks of the proposed PLL are shown and discussed below.

3.1. Phase Frequency Detector

The Phase Frequency Detector (PFD) mainly consists of two D-flip flops and NAND gates. The block diagram of the PFD is shown below in Figure 2. The main purpose of a PFD is to compare the phase and frequency of the input signal with the feedback signal. It has two output signals, UP and DOWN [7].

The design of the low power edge triggered D-Flip flop with 1.8 V is designed with a reduced number of transistors, and the respective output waveforms are shown in Figure 2. This D flipflop has been used in designing the PFD [8].

Figure 3 shows PFD with two input signals, reference signal CLKREF and the feedback signal which is the output of VCO, CLKVCO each as an input to the two D-flipflops. Output of the first D-flipflop enables the positive current source and another the negative current source of the charge pump. Assuming UP=DOWN=0, when CLKREF is high, UP rises thereby CLKVCO rises. When DOWN also is high, NAND gate resets both the D-flipflops. UP and DOWN are high at the same time for a short while, at this point the difference between their average values gives us the phase or frequency difference between both the input signals [9].

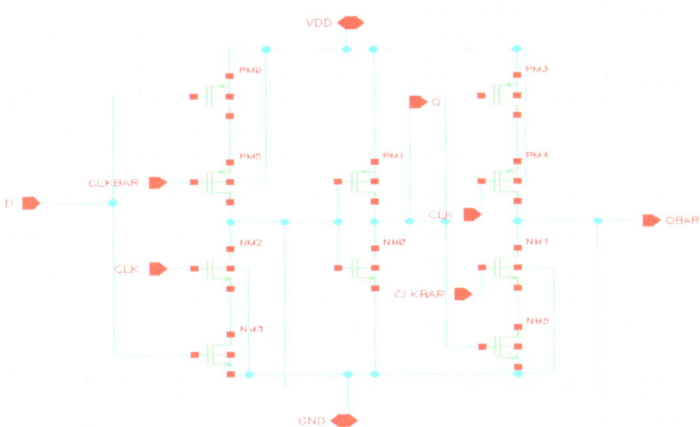

Figure 2. Schematic of D flip flop.

Figure 3. Schematic of Phase frequency detector.

3.2. Charge Pump and Low Pass Filter

The charge pump block in Figure 4 converts the output signals of PFD block i.e., UP and DOWN signals into a voltage that controls the VCO or $V_{control}$. This block also gives a constant voltage to hold the VCO at as the loop locks in at a particular frequency. When the UP signal is high, the positive current I_{PDI} flows through the circuit and increases the control voltage, When the downstream signal goes high, a negative I_{PDI} current flows through the circuit, which reduces the control voltage [10]. The output current of the charge pump is given below Equation (1):

$$I_{PDI} = K_{PDI} \times \Delta\emptyset \tag{1}$$

where,

$$K_{PDI} = \frac{I_{PUMP}}{2\pi}$$

$$\Delta\emptyset = \emptyset_{IN} - \emptyset_{REF}$$

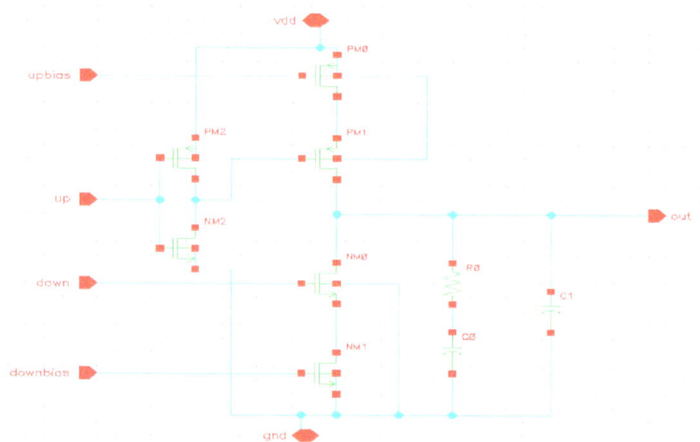

Figure 4. Schematic of charge pump with Low Pass Filter.

The control voltage of VCO is given in Equation (2):

$$V_{ctl}VCO = Kf \times I_{PDI} \tag{2}$$

The Low Pass Filter converts the charge pump current into voltage and the frequency of the VCO depends on the output of the LPF. When the charge pump current is positive, the oscillation frequency increases, otherwise it decreases [11]. Figure 4 shows the implementation of Charge Pump along with Low Pass Filter. Low Pass Filter here is important because it filters out higher frequencies and influences the speed of the circuit [12].

3.3. Voltage Controlled Oscillator

The implementation of a low current VCO circuit that is like a simple ring oscillator with an PMOS and NMOS transistor is done [13]. This limits the current that passes through each inverter, and is thus named a low current VCO. The transistors act as current source sand limit the current to the inverter and the inverter in the next stage is even more starved. This reduced the frequency due to reducing available current and increasing their time to charge and recharge [14]. The schematic of CSVCO is shown in Figure 5.

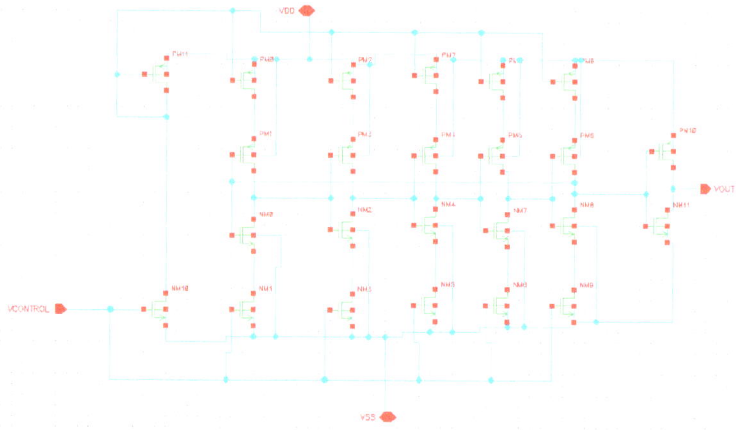

Figure 5. Schematic of Voltage controlled oscillator.

3.4. Frequency Divider

The output of the VCO is provided to the PFD via the frequency divider circuit. The frequency of the VCO output signal is divided by two by this frequency divider block [15]. Two D flipflops constructed using 5 NMOS and 5 PMOS transistors are used to implement the divider circuit, as shown in Figure 6.

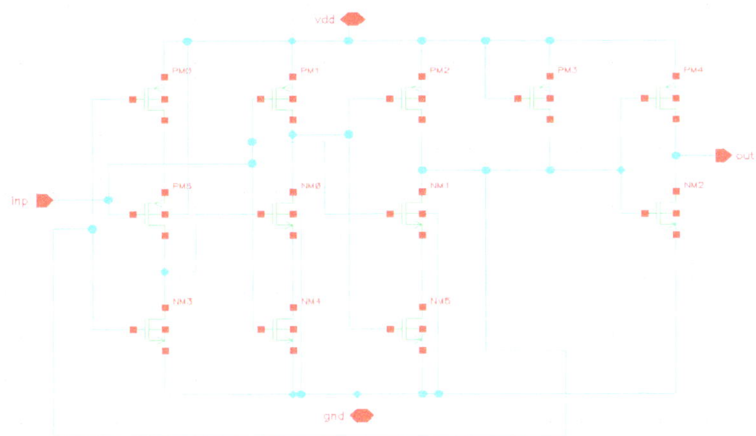

Figure 6. Schematic of the frequency divider.

4. Results and Observations

The design of a PLL is done in cadence virtuoso tool by Analog design environment using 90nm node. Here, the transient and DC analysis of proposed PLL and its blocks are discussed. A reference signal and a clock feedback signal are given as input to the PFD with output as UP and DOWN Signals. The output of the charge pump is given to the VCO, which acts as a Voltage control signal, and the output of VCO is given to a Frequency Divider where the output frequency is N/2, which is the feedback signal given back to PFD. Figure 7 shows the entire design of PLL schematic while Figure 8 shows the transient and DC analysis of the same.

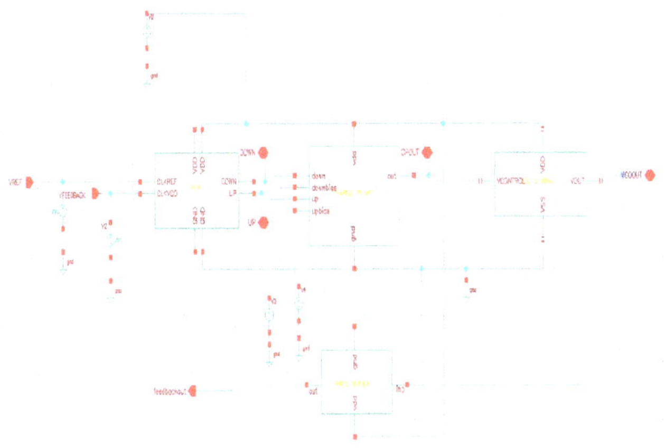

Figure 7. Architecture of Proposed PLL.

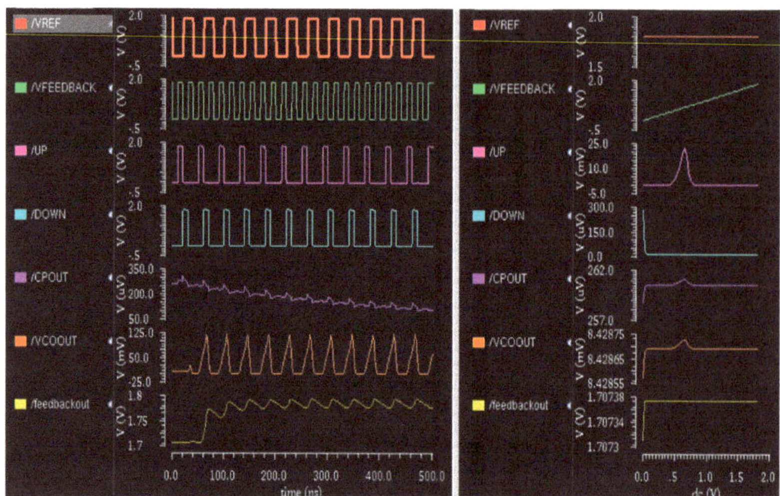

Figure 8. Transient analysis and DC analysis of proposed PLL.

Here, in Figure 7 the use of a novel design for PFD and a charge pump is used for reducing the power usage of whole system and it also shows how they are incorporated in the entire PLL.

Figure 8 shows transient and DC analysis of the proposed PLL that verifies the working of the PLL in an effective way. This not only shows the output PFD, i.e., UP and DOWN signals that act according to the clock signal but also show the output waveforms of the charge pump which varies according to the input signals. When both the input signals i.e., REF and FEEDBACK are high, the output of the charge pump is high; when both the input signals are low, there is a drop at the output signal of the charge pump, which is given as input to the VCO.

Figure 9 shows the two input signals to Charge Pump, UP and DOWN signals red and green respectively which are observed to be inverse of each other. The output signal of Charge Pump is in purple as OUT signal which is given as the input to the VCO as control signal.

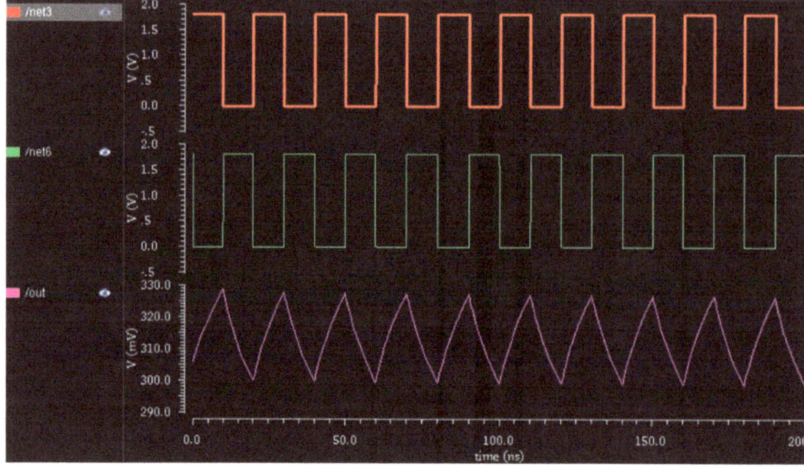

Figure 9. Output Waveforms of Charge Pump.

The output waveforms of VCO are shown in Figure 10 with respect to the input control signal. It shows the frequency of the signal produced by VCO at a constant voltage of $V_{control}$ = 800 mV.

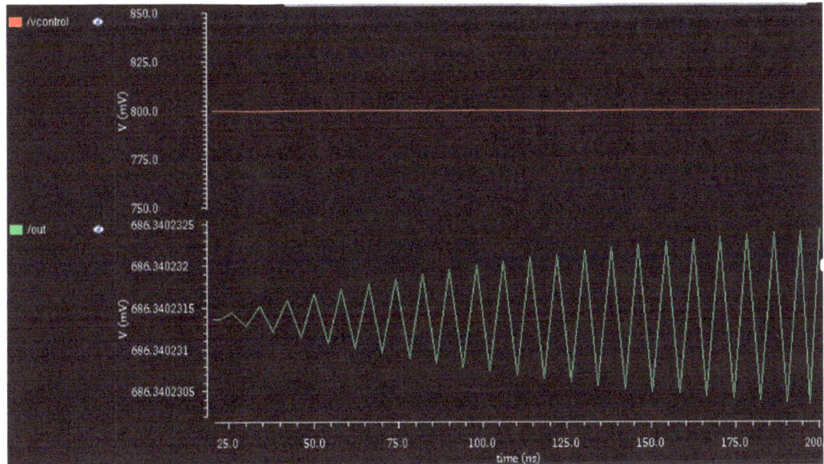

Figure 10. Output waveforms of VCO with $V_{control}$ signal.

Figure 11 shows input and output of the frequency divider. Visibly, the output pulses have half the frequency of the input pulses. In other words, the input frequency is divided by the frequency divider by a factor of 2.

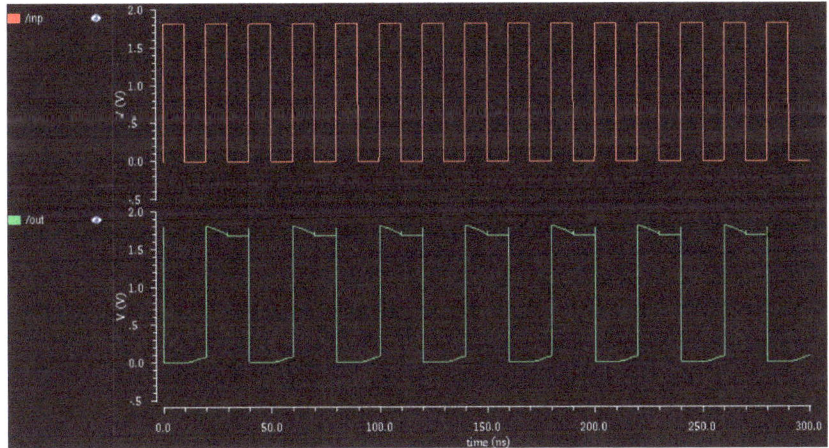

Figure 11. Output wave forms of Frequency Divider.

The DC total power analysis plot of the PLL is shown below Figure 12. The graph peaks at 194.24 µW which shows the total power consumption of the PLL circuit.

The results of the proposed work are better than the results of conventional state-of-the-art PLL design in terms of power consumption and the number of transistors used to design PLL, where power consumption is many times less than that of the conventional PLL. The comparative analysis of various parameters is presented in Table 1.

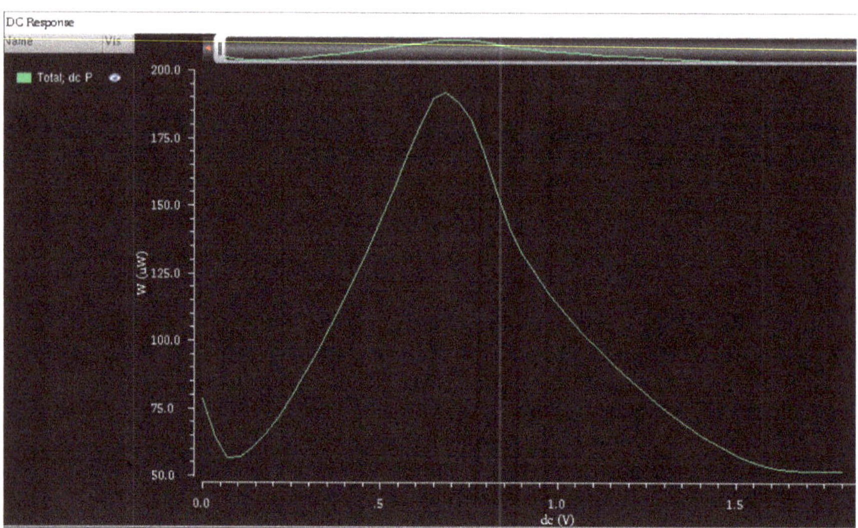

Figure 12. The plot of DC total power analysis.

Table 1. Parametric Analysis.

Parameters	Traditional PLL	Existing PLL	Proposed PLL
Number of transistors	112	56	48
V dd	-	1.8 V	1.8 V
Operating frequency	-	1 GHz	1 GHz
Power Consumption	-	4.2 mW	194.24 μW
Total Power Consumed	High	Medium	low
Technology	180 nm	90 nm	90 nm

Here, the comparative analysis of different parameters is shown. The parameters such as the number of transistors used in design of PLL, supply voltage, and operating frequency power consumed are used to implement PLL, and are analyzed above. From the analysis, the proposed PLL design has a 14% decrease in the number of transistors with a reduced area and 1000 times less power consumption. So, the proposed PLL can be effectively used in low power digital electronic applications and compact devices.

5. Conclusions

A design of PLL using a cadence virtuoso tool in an analog design environment by using GPDK 90nm technology with 1.8 V DC supply is performed. The simulation work presents a reduced number of transistors with a reduced area in proposed design with very low power consumed at DC voltage of 1.8 V. The Total Power Consumed by the proposed PLL design is 194.24micro-Watts. We know that the power consumed, the sizing of the transistors, and the selection of the power supply voltage at different levels may vary with the total power consumed, respectively. This not only allows the working of PLL at high speed, but also supports working at low power, which makes it very effective for low power applications.

Author Contributions: Conceptualization, C.K.P. and C.S.; methodology, N.P.S.; software, C.K.P.; validation, C.S. and M.R.K.; data curation, J.L.P.; writing—original draft preparation, C.K.P.; writing—review and editing, C.K.P., N.P.S., J.L.P., C.S. and M.R.K.; visualization, N.P.S.; supervision, C.S.; project administration, M.R.K. All authors have read and agreed to the published version of the manuscript.

Funding: This research received no external funding.

Institutional Review Board Statement: Not applicable.

Informed Consent Statement: Not applicable.

Data Availability Statement: Data can be shared on request.

Conflicts of Interest: The authors declare no conflict of interest.

References

1. Wang, H.; Momeni, O. A charge pump current mismatch compensation design for sub-sampling PLL. *IEEE Trans. Circuits Syst. II Express Briefs* **2021**, *68*, 1852–1856. [CrossRef]
2. Tsai, T.-H.; Sheen, R.-B.; Chang, C.-H.; Hsieh, K.C.-H.; Staszewski, R.B. A Hybrid-PLL (ADPLL/Charge-Pump PLL) Using Phase Realignment With 0.6-us Settling, 0.619-ps Integrated Jitter, and −240.5-dB FoM in 7-nm FinFET. *IEEE Solid-State Circuits Lett.* **2020**, *3*, 174–177. [CrossRef]
3. Chae, U.; Park, J.; Kim, J.-G.; Yu, H.-Y.; Cho, I.-J. CMOS voltage-controlled oscillator with high-performance MEMS tunable inductor. *Micro Nano Syst. Lett.* **2021**, *9*, 13. [CrossRef]
4. Bamigbade, A.; Khadkikar, V.; Hosani, M.A. A Type-3 PLL for Single-Phase Applications. *IEEE Trans. Ind. Appl.* **2020**, *56*, 5533–5542. [CrossRef]
5. Sánchez-Herrera, R.; Andújar, J.M.; Márquez, M.; Mejías, A.; Gómez-Ruiz, G. Self-Tuning PLL: A New, Easy, Fast and Highly Efficient Phase-Locked Loop Algorithm. *IEEE Trans. Energy Convers.* **2022**, *37*, 1164–1175. [CrossRef]
6. Patil, P.; Ingale, V. Design of a Low Power PLL in 90 nm CMOS Technology. In Proceedings of the IEEE 5th International Conference for Convergence in Technology, Bombay, India, 29–31 March 2019.
7. Mazza, G.; Panati, S. A Compact, Low Jitter, CMOS 65 nm 4.8–6 GHz Phase-Locked Loop for Applications in HEP Experiments Front-End Electronics. *IEEE Trans. Nucl. Sci.* **2018**, *65*, 1212–1217. [CrossRef]
8. Ghasemian, M.S.P.H. A straightforward quadrature signal generator for single-phase SOGI-PLL with low susceptibility to grid harmonics. *IEEE Trans. Ind. Electron.* **2021**, *69*, 6997–7007.
9. Razavi, B. Jitter-power trade-offs in PLLs. *IEEE Trans. Circuits Syst. I Regul. Pap.* **2021**, *68*, 1381–1387. [CrossRef]
10. Gong, H.; Wang, X.; Harnefors, L. Rethinking current controller design for PLL-synchronized VSCs in weak grids. *IEEE Trans. Power Electron.* **2021**, *37*, 1369–1381. [CrossRef]
11. Hoseinizadeh, S.M.; Ouni, S.; Karimi, H. Comparison of PLL-based and PLL-less vector current controllers. *IEEE J. Emerg. Sel. Top. Power Electron.* **2021**, *10*, 436–445. [CrossRef]
12. Achlerkar, P.D.; Panigrahi, B.K. New perspectives on stability of decoupled double synchronous reference frame PLL. *IEEE Trans. Power Electron.* **2021**, *37*, 285–302. [CrossRef]
13. Lai, W.-C. Performance Improved Voltage Controlled Oscillator and Double Balanced Mixer Chip Design. In Proceedings of the 2021 International Applied Computational Electromagnetics Society (ACES-China) Symposium, Chengdu, China, 28–31 July 2021; pp. 1–3.
14. Shettar, H.G.; Kotabagi, S.; Shanbhag, N.; Naik, S.; Bagali, R.; Nandavar, S. Frequency Multiplier using Phase-Locked Loop. In Proceedings of the IEEE 17th India Council International Conference (INDICON), New Delhi, India, 10–13 December 2020; pp. 1–5.
15. Adesina, N.O.; Srivastava, A.; Khan, A.U.; Xu, J. An Ultra-Low Power MOS2 Tunnel Field Effect Transistor PLL Design for IoT Applications. In Proceedings of the Electronics and Mechatronics Conference (IEMTRONICS), Toronto, ON, Canada, 21–24 April 2021; pp. 1–6.

Proceeding Paper

Advanced Driver Fatigue Detection by Integration of OpenCV DNN Module and Deep Learning [†]

Muzammil Parvez M. [1,*]**, Srinivas Allanki** [1,*]**, Govindaswamy Sudhagar** [2]**, Ernest Ravindran R. S.** [1]**, Chella Santosh** [1]**, Ali Baig Mohammed** [3] **and Mohd. Abdul Muqeet** [4]

[1] Department of Electronics and Communication Engineering, Koneru Lakshmaiah Education Foundation, Guntur 522502, India; ravindran.ernest@kluniversity.in (E.R.R.S.); csanthosh@kluniversity.in (C.S.)

[2] Department of Electronics and Communication Engineering, Bharath Institute of Higher Education and Research, Chennai 600126, India; sudhagar.ece@bharathuniv.ac.in

[3] School of Electronics and Communication Engineering, Reva University, Bangaluru 560064, India; alibaig.mohammad@reva.edu.in

[4] Department of Electrical Engineering, Muffakham Jah College of Engineering and Technology, Hyderabad 500034, India; ab.muqeet2013@gmail.com

* Correspondence: parvez190687@gmail.com (M.P.M.); aspvc02@gmail.com (S.A.)

† Presented at the International Conference on "Holography Meets Advanced Manufacturing", Online, 20–22 February 2023.

Abstract: Road safety is significantly impacted by drowsiness or weariness, which primarily contributes to auto accidents. If drowsy drivers are informed in advance, many fatal incidents can be avoided. Over the past 20 to 30 years, the number of road accidents and injuries in India has increased alarmingly. According to the experts, the main cause of this issue is that drivers who do not take frequent rests when travelling long distances run a great danger of becoming drowsy, which they frequently fail to identify early enough. Several drowsiness detection techniques track a driver's level of tiredness while they are operating a vehicle and alert them if they are not paying attention to the road. This study describes a noncontact way of determining a driver's tiredness utilising detecting techniques.

Keywords: OpenCV; DNN module; face detection; CNN

Citation: Parvez M., M.; Allanki, S.; Sudhagar, G.; R. S., E.R.; Santosh, C.; Mohammed, A.B.; Muqeet, M.A. Advanced Driver Fatigue Detection by Integration of OpenCV DNN Module and Deep Learning. *Eng. Proc.* **2023**, *34*, 15. https://doi.org/10.3390/HMAM2-14158

Academic Editor: Vijayakumar Anand

Published: 13 March 2023

1. Introduction

An automotive safety system called driver drowsiness detection works to stop accidents when the driver is about to nod off. According to several studies, weariness may play a role in up to 50% of specific roads and about 20% of all traffic accidents. Numerous auto accidents are significantly influenced by driver drowsiness. According to recent figures, collisions caused by driver drowsiness result in an estimated 1200 fatalities and 76,000 injuries per year [1]. A significant obstacle in the development of accident-avoidance systems is the detection or prevention of tiredness. A biological condition known as drowsiness or sleepiness occurs when the body is transitioning from an alert state to a sleeping state [2]. At this point, a motorist may become distracted and be unable to make decisions like avoiding crashes or using the brakes in a timely manner. There are clear indicators that a motorist may be drowsy, such as eye blinking/inability to keep eyes open, wobbling the head forward, and frequent yawning. Three categories can be used to categorize drowsiness detection are vehicle-based, behavioral-based and physiological based [3].

Vehicle-based measure: Several measurements include lane deviations, steering wheel movement, abrupt pressure on the accelerator pedal, etc. [4]. These parameters are continuously tracked, and if they fluctuate or cross a predetermined threshold, it means the driver is drowsy. Behaviour-based measurement: The metrics used here are based on the driver's actions, such as yawning, closing of the eyes, blinking, head posture, etc. A physiologically based measurement accesses the driver's conditions by attaching electrical devices to the skin. Electroencephalogram, Electrocardiogram, and Electrooculogram are examples of this.

Convolutional networks with three carefully chosen levels are used to suggest a new strategy for approximating the positions of important face points [5]. Two advantages exist: To discover every important spot, texture context data applied to the entire face is used initially. Second, the geometric relationships between key points are implicitly embedded in networks since they are adept at concurrently predicting all the critical points. The drawback is that while regionally sharing the weights of neurons on a single map enhances performance, global sharing of the weights does not perform well on images like faces or in some unusual poses or emotions.

Milla et al. [6] create a system that is not light-sensitive. For object detection, they employed the Haar algorithm, and for face classification, they used the OpenCV libraries' face classifier. Anthropometric parameters are used to determine the eye regions from the facial region. The amount of ocular closure is then determined by detecting the eyelid. The disadvantage is that this model performs poorly in dim lighting.

Omidyeganeh, M. et al. [7] used a camera to record face appearance. Face extraction from the image, eye detection, mouth detection, and alert creation are the four steps of the plan while asleep. Since a non-contact method is employed here, the system's shortcoming is that it depends on elements such as light, the camera, and other things.

Warwick et al. [8] demonstrated a wireless wearable sensor for drowsy driver detection systems. A drowsiness system is developed in two steps: the collection of physiological data using a biosensor and the identification of the essential components of drowsiness through analysis of the data. In the second stage, drowsy people are alerted via a mobile app and an algorithm that detects their tiredness. The heart and respiration rates of a driver are reliable indicators of weariness.

This approach describes a noncontact method for determining a driver's drowsiness by utilising OpenCV DNN module for detecting face and a convolution neural network model to classify his state.

2. Methodology

The entire process of the drowsiness detection system is carried out following image processing, which is a technique for carrying out some actions on an image. The traditional Drowsiness Detection System is depicted in Figure 1.

There are five steps in this system flow, and they basically consist of Acquiring video, detection of the face, detection of eyes and mouth, state assessment and at last categorisation into drowsy or non-drowsy. There are three fundamental steps in this approach. First, video is captured using the camera, then it is transformed into frames, then faces are identified, and last, deep learning is used to detect sleepiness. In comparison to contact-based methods, this technology is non-contact-based and will be inexpensive.

2.1. Acquiring Video

This uses the Kaggle Drowsiness dataset. This accessible dataset is meant to aid in research projects. The camera is used to capture the photographs. There are four classes in this dataset that are used to categorize the state of the individual. The dataset contains images with a resolution of 640×480 pixels.

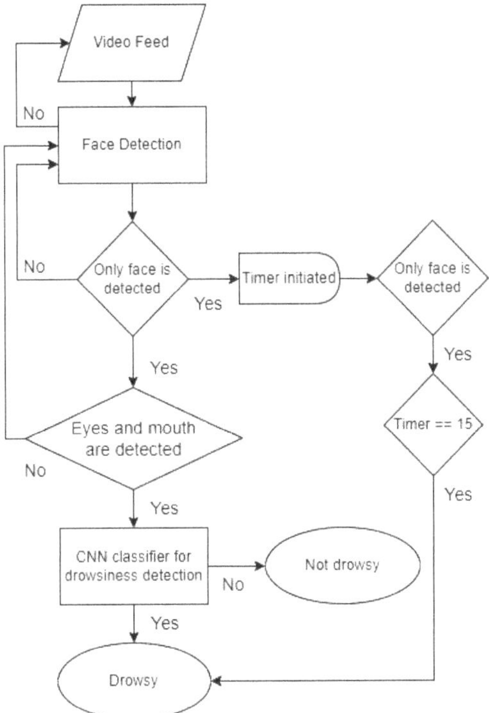

Figure 1. Proposed approach.

2.2. Detection of Face

Due to its simplicity, the Viola–Jones algorithm is frequently employed for face detection. Here, Haar characteristics are retrieved to identify facial, bodily, skeletal, and other markers [ ]. As illustrated in Figure 2, Edge, line, and four-sidedness are the three Haar-like properties. The Haar-like feature needs to provide you with a higher score because our faces have complicated forms with darker and brighter areas. Computation of Haar features comprises comparing pixel intensities within predefined rectangular regions which may be either non-negative (white) or negative (black).

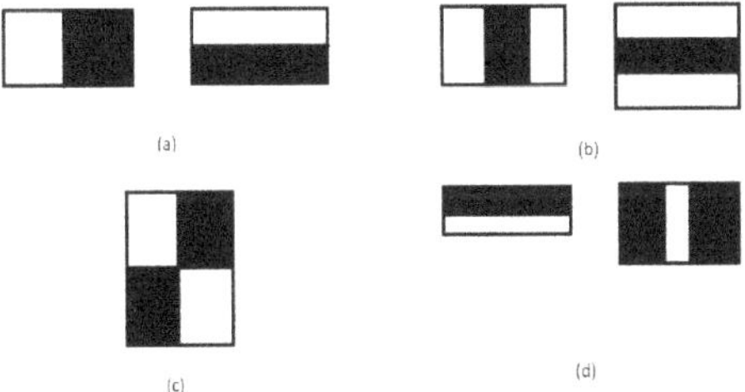

Figure 2. Features of Haar (**a**) Edge (**b**) Line (**c**) Four rectangles (**d**) Face detection [ ].

For that integral picture calculates the sum of pixel values in an image or rectangular section of an image, which helps us accomplish these laborious computations fast and effectively. Even after using the integral image approach, each image sub-window still has more than 180,000 rectangle features. Even though it is very efficient to compute each feature individually, doing it for the entire collection of features is expensive.

To find the optimal features, the AdaBoost method is employed [10]. In the end, the algorithm creates a simple standard to determine if a property is considered to be valuable.

Another choice is to use the "Dlib" library for face detection. Dlib is an open-source platform for machine learning applications. The 68 coordinates (x, y) that map the facial locations on a person's face are estimated using the Dlib. It is a facial detector with trained models for landmarks. The pre-trained model using the iBUG300-W dataset was used to locate these spots.

The multi-task cascaded convolutional network (MTCNN) model is an additional technique for face detection. This model [11] can retain real-time speed while outperforming several face-detection benchmarks. Three convolutional networks are present (P-Net, R-Net, and O-Net). High accuracy is possible when a deep neural network is used. Using three networks, each with many layers allows for more precision because each network can adjust the results of the one that came before it. This model also locates both large and small faces using an image pyramid. NMS, R-Net, and O-Net all assist in the rejection of several inaccurate bounding boxes despite the potentially overwhelming amount of data generated.

The most recent version of OpenCV has a deep neural network (DNN) module with an excellent pre-trained convolutional neural network (CNN) for face detection. The new model performs better in face detection when compared to earlier models like Haar. It is a Caffe model based on the single-shot multibox detector with ResNet-10 architecture at its heart (SSD).

2.3. Drowsiness Detection:

The system's classification procedure involves drowsiness detection. For classification, many machine-learning techniques have previously been created. You can choose one of two classification kinds. (i) Determine if the mouth or eyes are open or closed by analysing the eye and mouth region of interest. (ii) Examine the full area of the face that is of interest. Here, using a modified convolution neural network [12], a novel training approach is developed. CNNs are crucial resources for deep learning and are particularly well-suited for studying picture data.

Convolutional networks are composed of input, output, and perhaps one or more hidden layers as shown in Figure 3. In contrast to a regular neural network, a convolutional network's layers include neurons organized in three dimensions (width, height, and depth dimensions). On account of this, CNN can transform a three-dimensional input volume into an output volume. The hidden layers are composed of convolution layers, pooling layers, normalizing layers, and completely.

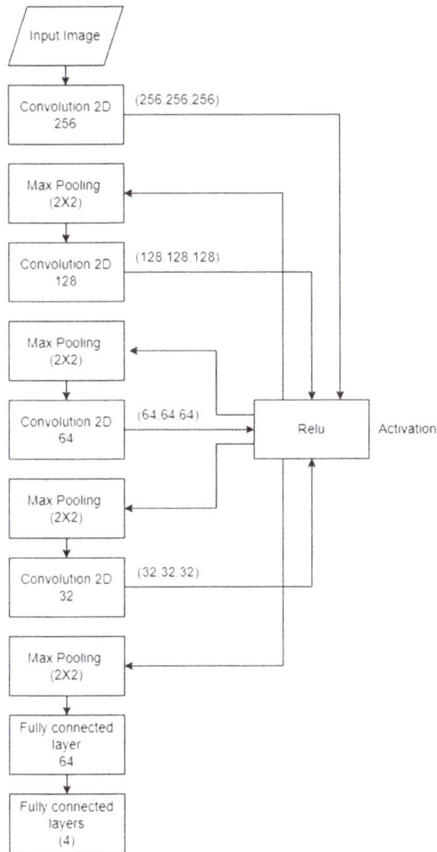

Figure 3. Modified CNN Model.

3. Result

In addition to other research projects, this effort requires a variety of deep learning techniques. The first goal of this effort is face detection with occlusions. There were four implemented algorithms on Figure 4. Red squares are utilized to provide a visual representation of the outcomes generated by the algorithms.

Figure 4. Test image [13].

Although the Viola–Jones algorithm executes relatively quickly, it cannot recognize side faces and produces a greater number of false positives as shown in Figure 5.

Figure 5. Face detection using Viola–Jones [13].

Dlib cannot distinguish between small faces. Dlib's facial detector had an issue because it cannot identify faces smaller than 80×80 pixels. Since the images were so small, the faces were even smaller, as illustrated in Figure 6. Because the face size cannot be very small and up-sampling the image may lengthen processing time, the image was scaled up by a factor of 2 for testing. However, this is a major issue when using Dlib.

Figure 6. Face detection using Dlib [13].

MTCNN slightly outperformed Dlib in terms of outcomes, although Dlib cannot recognise very small faces. Additionally, MTCNN may produce the greatest results if the size of the images is sufficiently large and adequate illumination, minimal occlusion, and primarily front faces are guaranteed, as shown in Figure 7.

The Caffe model from OpenCV used by the DNN module is the best. It is also a helpful parameter in the sleepiness detection approach that has been developed, and as shown in Figure 8, it functions well with occlusion, fast head movements, and the ability to recognize side faces.

Figure 7. Face detection using MTCNN [13].

Figure 8. Face detection using DNN [13].

The results are acquired using an AMD Ryzen 7 5800H processor and an Nvidia RTX 3050 GPU, and the images passed are all 640 × 360 pixels, with the exception of the DNN module, which is still receiving 300 × 300 pixels as usual. Frame rate [14] for different methods is illustrated in the Table 1.

Table 1. Frame rate.

Methods	Frame Rate
Haar	38
Dlib	32
MTCNN	34
DNN	46

Figure 9 the model performance during training process. The red line indicates the validation accuracy per epoch and the blue line indicates training accuracy per epoch of the model.

Table 2 illustrates how the suggested system performs better than the current drowsiness detection techniques [15,16].

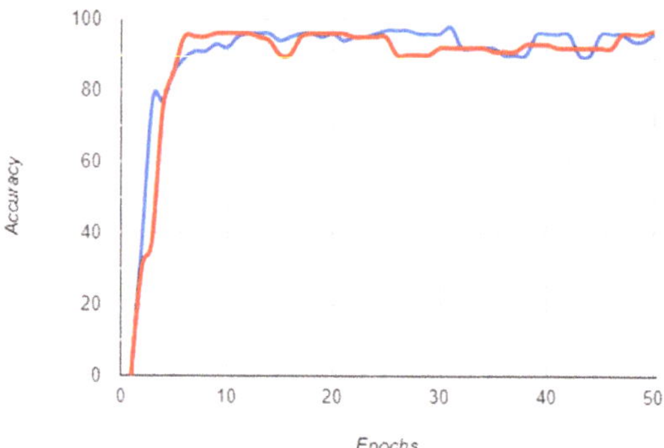

Figure 9. Model performance after 50 epochs of training.

Table 2. Examining Performance.

Algorithm	Accuracy
Picot [17]	82.1%
Akrout [18]	90.2%
Dricare [19]	93.6%
Zhang [20]	85.9%
Proposed Algorithm	96.8%

4. Conclusions

This approach will go through a method for determining a driver's level of intoxication in this section. For face detection, it was shown that the OpenCV DNN module outperformed Viola–Jones, Dlib, and MTCNN. The categorization process employs a modified CNN. The accuracy rate of the system is 96.8%. By evaluating the validation accuracy while using the validation dataset for model training, the model was proven to be accurate. In the future, the detection can be improved by using an infrared camera for low-light situations. Counting the frequency of yawns over a certain amount of time might also help identify drowsiness. Further, one can utilize a multi-model machine-learning approach and include additional modalities, such as the audio channel, in addition to the video frames, to enhance performance.

Author Contributions: All authors contributed to the study conception and design. Data collection and analysis were performed by M.P.M., S.A., G.S., E.R.R.S., C.S., A.B.M. and M.A.M. The first draft of the manuscript was written by S.A. and M.P.M. and all authors commented on previous versions of the manuscript. All authors have read and agreed to the published version of the manuscript.

Funding: The authors declare that no funds, grants, or other support were received during the preparation of this manuscript.

Institutional Review Board Statement: Not applicable.

Informed Consent Statement: Not applicable.

Data Availability Statement: Research Data sets for this submission was used from (https://www.kaggle.com/datasets/dheerajperumandla/drowsiness-dataset, accessed on 20 November 2022).

Conflicts of Interest: The authors declare that they have no known competing financial interest or personal relationship that could have appeared to influence the work reported in this paper.

References

1. Dhawde, P.; Nagare, P.; Sadigale, K.; Sawant, D.; Mahajan, J.R. Drowsiness Detection System. *Int. J. Eng. Res. Technol. (IJERT) Iconect* **2015**, *3*, 1–2.
2. Lal, S.K.; Craig, A. A critical review of the psychophysiology of driver fatigue. *Biol. Psychol.* **2001**, *55*, 173–194. [CrossRef] [PubMed]
3. Ramzan, M.; Khan, H.U.; Awan, S.M.; Ismail, A.; Ilyas, M.; Mahmood, A. A Survey on State-of-the-Art Drowsiness Detection Techniques. *IEEE Access* **2019**, *7*, 61904–61919. [CrossRef]
4. Sahayadhas, A.; Sundaraj, K.; Murugappan, M. Detecting Driver Drowsiness Based on Sensors: A Review. *Sensors* **2012**, *12*, 16937–16953. [CrossRef]
5. Sun, Y.; Chen, Y.; Wang, X.; Tang, X. Deep learning face representation by joint identification-verification. *Adv. Neural Inf. Process. Syst.* **2014**.
6. Alshaqaqi, B.; Baquhaizel, A.S.; Ouis, M.E.A.; Boumehed, M.; Ouamri, A.; Keche, M. Driver Drowsiness Detection System. In Proceedings of the 2013 8th International Workshop on Systems, Signal Processing and their Applications (WoSSPA), Algiers, Algeria, 12–15 May 2013; IEEE: Piscataway, NJ, USA, 2013; pp. 151–155.
7. Omidyeganeh, M.; Javadtalab, A.; Shirmohammadi, S. Intelligent driver drowsiness detection through fusion of yawning and eye closure. In Proceedings of the 2011 IEEE International Conference on Virtual Environments, Human-Computer Interfaces and Measurement Systems Proceedings, Ottawa, ON, Canada, 19–21 September 2011; IEEE: Piscataway, NJ, USA, 2011; pp. 1–6.
8. Warwick, B.; Symons, N.; Chen, X.; Xiong, K. Detecting driver drowsiness using wireless wearables. In Proceedings of the 2015 IEEE 12th International Conference on Mobile ad Hoc and Sensor Systems, Dallas, TX, USA, 19–22 October 2015; IEEE: Piscataway, NJ, USA, 2015; pp. 585–588.
9. Zhou, W.; Chen, Y.; Liang, S. Sparse Haar-Like Feature and Image Similarity-Based Detection Algorithm for Circular Hole of Engine Cylinder Head. *Appl. Sci.* **2018**, *8*, 2006. [CrossRef]
10. Schapire, R.E. Explaining AdaBoost. In *Empirical Inference*; Schölkopf, B., Luo, Z., Vovk, V., Eds.; Springer: Heidelberg, Germany, 2013. [CrossRef]
11. Zhang, K.; Zhang, Z.; Li, Z.; Qiao, Y. Joint Face Detection and Alignment Using Multitask Cascaded Convolutional Networks. *IEEE Signal Process. Lett.* **2016**, *23*, 1499–1503. [CrossRef]
12. Suresh, M.; Sinha, A.; Aneesh, R.P. Real-Time Hand Gesture Recognition Using Deep Learning. *Int. J. Innov. Implement. Eng.* **2019**, *1*, 11–15.
13. Mossholder, T. Photo by Tim Mossholder on Unsplash, Beautiful Free Images & Pictures. 2019. Available online: https://unsplash.com/photos/hOF1bWoet_Q (accessed on 15 February 2023).
14. Danelljan, M.; Häger, G.; Khan, F.S.; Felsberg, M. Discriminative Scale Space Tracking. *IEEE Trans. Pattern Anal. Mach. Intell.* **2017**, *39*, 1561–1575. [CrossRef] [PubMed]
15. Saini, V.; Saini, R. Driver drowsiness detection system and techniques: A review. *Int. J. Comput. Sci. Inf. Technol.* **2014**, *5*, 4245–4249.
16. Albadawi, Y.; Takruri, M.; Awad, M. A Review of Recent Developments in Driver Drowsiness Detection Systems. *Sensors* **2022**, *22*, 2069. [CrossRef] [PubMed]
17. Picot, A.; Charbonnier, S.; Caplier, A.; Vu, N.-S. Using retina modelling to characterize blinking: Comparison between EOG and video analysis. *Mach. Vis. Appl.* **2011**, *23*, 1195–1208. [CrossRef]
18. Deng, W.; Wu, R. Real-Time Driver-Drowsiness Detection System Using Facial Features. *IEEE Access* **2019**, *7*, 118727–118738. [CrossRef]
19. Akrout, B.; Mahdi, W. Spatio-temporal features for the automatic control of driver drowsiness state and lack of concentration. *Mach. Vis. Appl.* **2015**, *26*, 1–13. [CrossRef]
20. Zhang, Y.; Hua, C. Driver fatigue recognition based on facial expression analysis using local binary patterns. *Optik* **2015**, *126*, 4501–4505. [CrossRef]

Proceeding Paper

Light Sheet Fluorescence Microscopy Using Incoherent Light Detection [†]

Mariana Potcoava [1,*], Christopher Mann [2,3], Jonathan Art [1] and Simon Alford [1]

[1] Department of Anatomy and Cell Biology, University of Illinois at Chicago, 808 South Wood Street, Chicago, IL 60612, USA; jart@uic.edu (J.A.); sta@uic.edu (S.A.)

[2] Department of Applied Physics and Materials Science, Northern Arizona University, Flagstaff, AZ 86011, USA; christopher.mann@nau.edu

[3] Center for Materials Interfaces in Research and Development, Northern Arizona University, Flagstaff, AZ 86011, USA

* Correspondence: mpotcoav@uic.edu; Tel.: +1-312-355-0328

† Presented at the International Conference on "Holography Meets Advanced Manufacturing", Online, 20–22 February 2023.

Abstract: We previously developed an incoherent holography technique for use in lattice light sheet (LLS) microscopes that represents a specialized adaptation of light sheet microscopy. Light sheet instruments resolve 3D information by illuminating the sample at 90° to the imaging plane with a sheet of laser light that excites fluorophores in the sample only in a narrow plane. Imaging this plane and then moving it in the imaging z-axis allows construction of the sample volume. Among these types of instruments, LLS microscopy gives higher z-axis resolution and tissue depth penetration. It has a similar working principle to light sheet fluorescence microscopy but uses a lattice configuration of Bessel beams instead of Gaussian beams. Our incoherent light detection technique replaces the glass tube lens of the original LLS with a dual diffractive lens system to retrieve the axial depth of the sample. Here, we show that the system is applicable to all light sheet instruments. To make a direct comparison in the same emission light path, we can imitate the nature of non-Bessel light sheet systems by altering the mask annuli used to create Bessel beams in the LLS system. We change the diffractive mask annuli from a higher NA anulus to a smaller NA anulus. This generates a Gaussian excitation beam similar to conventional light sheet systems. Using this approach, we propose an incoherent light detection system for light sheet 3D imaging by choosing a variable NA and moving only the light sheet while keeping the sample stage and detection microscope objective stationary.

Keywords: incoherent holography; light sheet; fluorescent microscopy

Citation: Potcoava, M.; Mann, C.; Art, J.; Alford, S. Light Sheet Fluorescence Microscopy Using Incoherent Light Detection. *Eng. Proc.* **2023**, *34*, 16. https://doi.org/10.3390/HMAM2-14156

Academic Editor: Vijayakumar Anand

Published: 13 March 2023

1. Introduction

Continuous irradiation causes photodamage and phototoxicity when imaging living samples and specimens. In order to overcome this problem, light sheet fluorescence microscopy (LSFM) was developed as an imaging technique with good optical sectioning, to record faster and scan larger sample volumes at low radiation intensities over longer time frames [1]. The LSFM is similar to laser scanning confocal microscopy (LSCM) [2], but the emitted fluorescence light is collected by the detection objective in a perpendicular direction from the excitation light and without the need for a confocal aperture. In LSFM, the excitation light does not pass through the entire sample; instead, the sample is illuminated from the side with a thin light sheet (LS) beam. This clever idea allows the LSFM to collect the emitted light from fluorophores localized in the thin LS plane only and not from fluorophores belonging outside the LS plane. This results in a low light dose usage, and therefore reduces both photobleaching and phototoxicity [3–5]. LSFM is also referred to as selective-plane illumination microscopy (SPIM) when imaging large samples [1]. The axial resolution is given by both the thickness of the LS (ideally it is extremely thin) and the detection NA [6,7]. If we choose a microscope objective with an NA of 1.1, we can obtain a very thin light sheet. However, the light sheet confocal length, over which the sheet is thin, must be matched to the FOV of the recorded images or the size of the samples. The thinner the sheet (and higher the NA), the shorter the thin region is.

In most of the cases, the light sheet has a Gaussian profile at the specimen due to the Gaussian shape of the laser beams. A thicker Gaussian light sheet can cover a larger FOV but has a lower resolution and optical sectioning. Non-Gaussian beams using arrays of Bessel beams, called "lattice" beams, have been employed to generate images with excellent scanning efficiency and resolution, by using very thin light sheets. The lattice light sheet microscopy (LLSM) system uses the LSFM principle with an optical lattice created with Bessel beams, resulting in a light sheet with a thickness of ~400 nm [8]. The typical acquisition rate of an LLSM system is hundreds of frames per second, making LLSM the ultimate tool for live-cell fluorescence imaging.

The lattice light sheet, Figure 1a, is formed by superimposing a linearly polarized sheet of light with a binary phase map of Bessel beams on a binary spatial light modulator (SLM), which is conjugated to the image plane of the excitation objective. Before reaching the sample plane, the beam passes a Fraunhofer lens and is further projected onto a transparent optical annulus, Figure 1b, to eliminate unwanted diffraction orders and lengthen the light sheet. In dithered mode, a 2D lattice pattern is oscillated in the x-axis using a galvanometer (x-galvo), to create a uniform sheet, while another galvanometer mirror moves together with the detection objective, in the z-axis (z-galvo), to scan the sample volume.

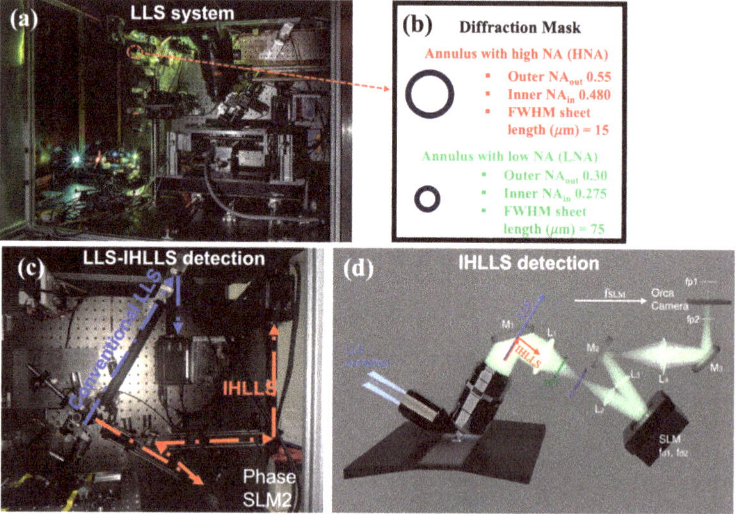

Figure 1. (**a**) The LLS instrument [8]; (**b**) the diffraction annular mask filters of inner diameter, NA_{in}, and outer diameter, NA_{out}, to filter out undiffracted light and unwanted higher diffraction orders; (**c**) the detection arms for the LLS instrument with the IHLLS detection arm incorporated; and (**d**) optical schematics of IHLLS. The LLS and IHLLS with HNA use higher NAout and a shorter light sheet (more like Bessel beams), here NAout = 0.55, NAin = 0.48, and sheet length 15 μm. The LLS and IHLLS with LNA use lower NAout and longer light sheet (more like Gaussian beams), here NAout = 0.3, NAin = 0.275, and sheet length 75 μm. The detection magnification Mdet = 62.5 and the illumination wavelength $\lambda_{excitationn}$ = 488 nm. The LLS, or IHLLS 1L with one diffractive lens, f_{SLM} = 400 mm, are used for calibration purposes. The IHLLS 2L with two diffractive lenses, f_{d1} = 220 mm and f_{d2} = 2356 mm, are used for incoherent imaging. Lenses L_1=L_4 with focal lengths 175 mm, L_2=L_3 with focal lengths 100 mm; mirrors M_1, M_2, M_3; f_{p1} and f_{p2} are the imaging positions of the two diffractive lenses, f_{d1} and f_{d2}.

A more recent approach of the LLSM is the new detection path, Figure 1c,d, that uses the incoherent holography idea, called an incoherent holographic lattice light sheet (IHLLS) [9], to scan the sample volume without moving the detection objective. It is possible by projecting two diffractive lenses at various z-galvo positions, using a phase SLM, to encode the depth position of the sample being imaged. The annular mask, Figure 1b, mentioned above, was centered on an anulus with a higher NA (HNA), NAout = 0.5, NAin = 0.485, and a sheet length 15 μm; therefore, the beams used for excitation were more Bessel-like beams. The intensity profile of the obtained optical lattice is determined by both the applied binary phase map and the geometry of the transparent annulus.

Here, we propose to expand our previous IHLLS detection technique that uses a HNA by choosing a diffractive mask annulus with a lower NA (LNA). In this case, the anulus has the following parameters: NAout = 0.3, NAin = 0.275, and sheet length 75 μm. The new detection approach uses a Gaussian-like excitation beam similar to conventional light sheet systems. This provides a tradeoff between the size of the imaged beam and the resolution of the ultimate image in the z-plane. This has advantages for imaging objects at larger intercellular scales.

2. Imaging Performances and Diffraction Mask Parameter Selection

In LLS, near-non-diffracting Bessel beams are created at the focal position of the excitation objective, by projecting an annular excitation beam, Figure 2a, in the back focal plane of the excitation objective. While the z-galvo and z-piezo are moved along the z axis to acquire stacks in LLS/IHLLS 1L, Figure 2b, in IHLLS 2L only the z-galvo is moved at various z positions, Figure 2c. The imaging area/volume is limited to a small FOV, Figure 2d red square, due to the tradeoff between the sheet's propagation length and its thickness. Although IHLLS 1L is the incoherent version of the LLS detection module, the axial resolution in LLS is better than the axial resolution in IHLLS 1L due to the blurry effect of the diffractive lens that focusses to infinity. The FOV limitation is no longer an issue in IHLLS 2L because the combination of the z-galvo motion in the range -40 μm to $+40$ μm and the dual diffractive lenses uploaded onto the phase SLM allows scanning for the whole FOV of the CMOS detector, which is 208×208 μm^2. Moreover, the axial resolution is better in IHLLS 2L due to the fact that the focus position is found numerically.

The NA of the beam is an overall combination of the inner and outer numerical apertures (NAs) of the annular mask filter, NA$_{\text{out}}$, and NA$_{\text{in}}$. The Bessel-like beams are considered to be created using those annuli on the diffraction mask with NA$_{\text{out}} \geq 0.5$ and the Gaussian-like beams created with the annuli, with NA$_{\text{out}} < 0.4$. The thickness of the pattern generated by these beams at the focal position of the excitation objective is given by $w_{\text{sheet}} = \lambda_{\text{excitation}}/2\text{NA}_{\text{out}}$, and the sheet length or the FOV by $\frac{\lambda_{\text{excitation}}}{n(\cos\theta_{\text{in}} - \cos\theta_{\text{out}})}$, where $\theta_{\text{in}} = \text{arsin}(\text{NA}_{\text{in}}/n)$, $\theta_{\text{out}} = \text{arsin}(\text{NA}_{\text{out}}/n)$, and n = 1.33 [7]. Although increasing the NA difference produces a thinner light sheet illumination profile, which provides higher axial resolution, the length that the light sheet spans is reduced, and thus it is hard to further compress the light sheet thickness while maintaining a relatively large FOV. The sheet parameters of the two diffraction mask filters at 488 nm are $w_{\text{sheet-HNA}} = 443.6$ nm, $w_{\text{sheet-LNA}} = 813.3$ nm, and the FWHM sheet lengths are $\text{FWHM}_{\text{sheet-HNA}} = 15$ μm and $\text{FWHM}_{\text{sheet-LNA}} = 75$ μm.

Imaging can be done with various sheet lengths: cultured cells and intracellular events could be imaged using light sheets with lengths smaller than ~20 μm, small organoids with sheets of lengths ~30 to 40 μm, and large organoids with sheets of lengths ~50 to 70 μm.

The beams generated by using the two annuli are shown in Figure 3. The upper row shows the optical lattice generated with the HNA annulus and the bottom row the optical lattice generated by the LNA annulus. Optical lattices can be designed to balance the confinement and the thickness by adjusting the spacing between each pair of beams. As an example, the lattice in Figure 3a is designed as a coherent superposition of 30 Bessel beams with a spacing of 0.99. Similarly, the lattice in Figure 3b is designed by using 26 Bessel beams with a spacing of 1.755. A single Bessel beam is shown in Figure 3c,d, respectively.

We can see that the lower NA beam has a Gaussian-like shape. The beams at the sample plane, corresponding to the patterns in Figure 3a,b, are shown in Figure 3e,f, and the same beams but with the dithered mode are shown in Figure 3g,h.

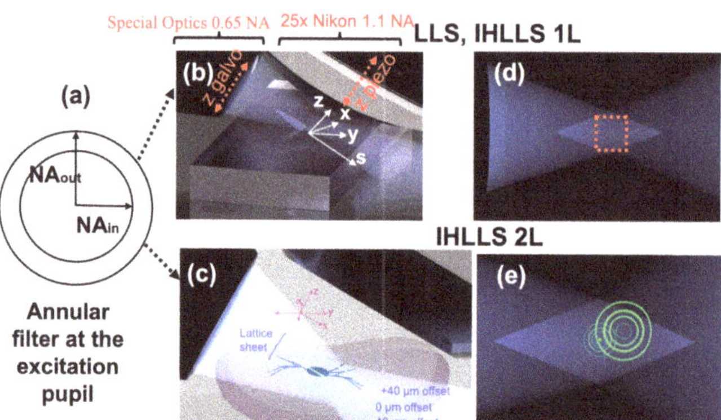

Figure 2. Excitation and scanning geometry in LLS, IHLLS Systems. (**a**) The diffraction mask is conjugated with the BFP of the excitation objective; (**b**) the scanning geometry in LLS and IHLLS 1L, the z−galvo and z−piezo are moved along the z axis to acquire imaging stacks; (**c**) The scanning geometry in IHLLS 2L; the z-galvo is moved at various z positions but the z-piezo is kept fixed; (**d**) The FOV in a conventional LLS system or the incoherent version with one diffractive lens, with dithering scanning modality and HNA, is about $52 \times 52\ \mu m^2$ (red area); (**e**) the diffractive lens effect in IHLLS 2L system. Excitation objective: Special Optics 0.65 NA, 3.74 mm working water dipping lens; detection objective: Nikon CFI Apo LWD 25×—water dipping, 1.1 NA, 3 mm working distance.

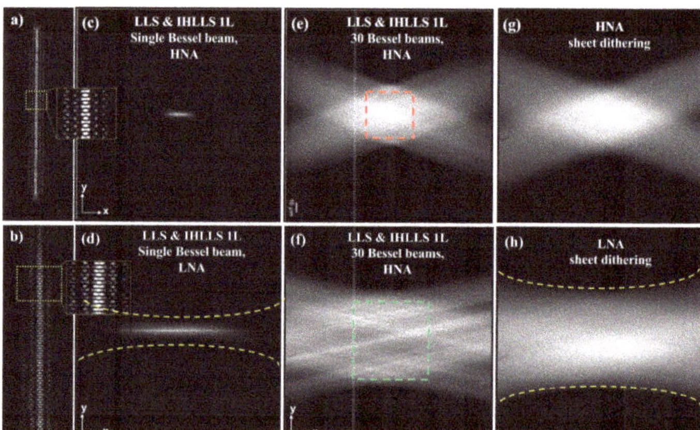

Figure 3. Optical lattice generation using HNA and LNA annuli; (**a**) HNA, 30 Bessel beams optical lattice with 0.99 spacing; the width of the light sheet in the center of the FOV is about 400 nm; (**b**) LNA, 26 Bessel beams optical lattice with 1.755 spacing; the width of the light sheet in the center of the FOV is about 800 nm; (**c**,**d**) Single Bessel beam at the sample plane; and (**e**,**f**) the Bessel beams corresponding to the optical lattices from (**a**,**b**) at the sample plane. The scanning area in a conventional LLS (HNA) is at best $52 \times 52\ \mu m^2$ (red square in (**e**)), and in a conventional LLS (LNA) about $78 \times 78\ \mu m^2$ (green square in (**f**)); (**g**,**h**) The Bessel beams corresponding to the optical lattices from (**a**,**b**) with the dithered mode at the sample plane.

3. Results

To examine the effects of applying IHLLS holography with LNA, we performed three experiments for this study.

3.1. LLS and IHLLS 1L Imaging with LNA

The first was done using the conventional LLS pathway where the z-galvo was stepped in $\delta z = 0.101$ μm increments through the focal plane of a 25× Nikon objective, for a displacement of $\Delta z = 40$ μm, Figure 4a. The detection objective was simultaneously moved the same distance with a z-piezo controller. The second set of images was obtained using the IHLLS 1L with a focal length $f_{SLM} = 400$ mm displayed on the SLM, where both the z-galvo and z-piezo were again stepped with the same $\delta z = 0.101$ μm increments through the focal plane of the objective for the same displacement $\Delta z = 40$ μm, Figure 4b. The scanning area in a conventional LLS with HNA is at best 52×52 μm^2 or 78×78 μm^2 in a conventional LLS with LNA (green dashed rectangles). Therefore, to enlarge the scanned region these ROI areas can be moved in a mosaic fashion by moving the sample.

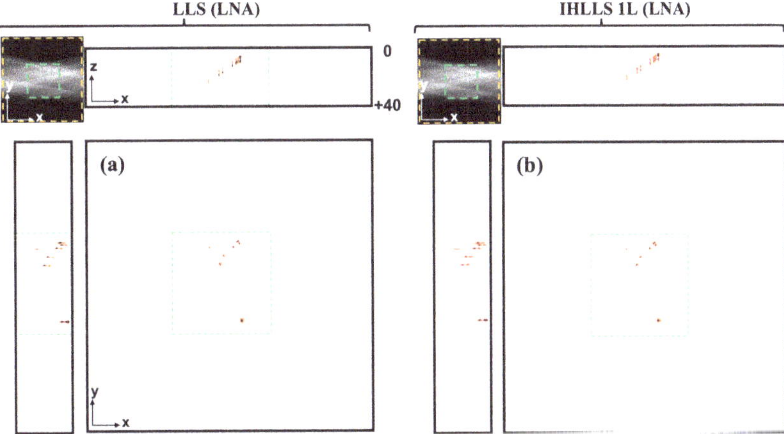

Figure 4. Tomographic imaging of 0.5 μm polystyrene beads, maximum FOV 208×208 μm^2, in a conventional LLS (**a**) and IHLLS 1L of focal length 400 nm (**b**), without deconvolution, with LNA. On the sides and above are shown the max projections through the volume (400 z-galvo steps). The Bessel beams are displayed in the upper left corner of each xy-projection to show the orientation of the beams, FOV 208×208 μm^2. The area enclosed inside the colored dashed rectangles, in the corner insets of (**a**,**b**), are as follows: green—the scanning area for the original LLS with LNA, 78×78 μm^2, and yellow—the maximum FOV scanning area for the IHLLS 2L. We notice that LLS and IHLLS 1L cannot achieve the maximum FOV. The maximum FOV is achieved by using the IHLLS 2L, as showing in the next section.

3.2. IHLLS 2L Imaging

Our incoherent system enables full complex-amplitude modulation of the emitted light for extended FOV and depth. The second set of images was obtained using the IHLLS pathway with two super-imposed diffractive lenses displayed on the SLM comprising randomly selected pixels (IHLLS 2L), where only the z-galvo was moved within the same $\Delta z = 40$ μm displacement range, above and below the reference focus position of the objective (which corresponds to the middle of the camera FOV), at z_galvo = ±40 μm, ±30 μm, ±20 μm, ±10 μm, and 0 μm, Figure 5. The two wavefronts interfere with each other at the camera plane, to create Fresnel holograms. Four interference patterns were created using a phase shifting technique ($\theta = 0$, $\theta = \pi/2$, $\theta = \pi$, $\theta = 3\pi/2$) and further combined mathematically to obtain the complex amplitude of the object point at the camera plane.

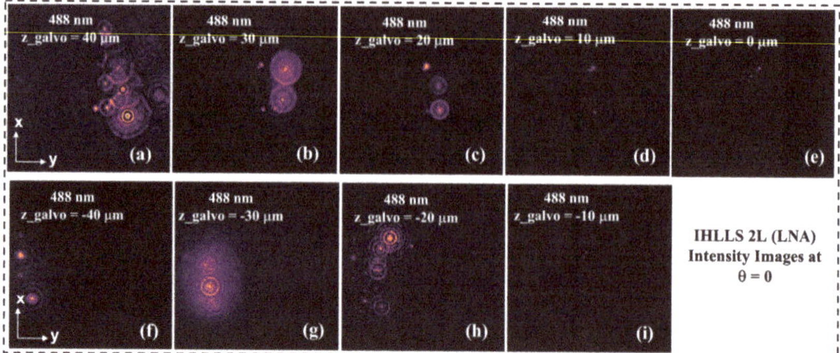

Figure 5. IHLLS 2L 500 nm beads holography, intensity images at phase shift, θ = 0; z-galvo position: (**a**) 40 μm, (**b**) 30 μm, (**c**) 20 μm, (**d**) 10 μm, (**e**) 0 μm, (**f**) −40 μm, (**g**) −30 μm, (**h**) −20 μm, and (**i**) −10 μm.

The max projection of all z-planes where the beads were found are displayed in Figure 6a. They show the complex holograms propagated to the best focal plane. We observed that IHLLS 2L performed better in comparison to the conventional LLS system in dithering mode regarding the scanning/detection area. The scanned area with detected beads in an IHLLS 2L system with LNA is the full FOV of the CMOS detector while the scanned area with detected beads in a LLS system with LNA is about 78×78 μm^2. In terms of axial resolution, we already showed [9] that the axial resolution of IHLLS 2L was higher than that of LLS because the objects are localized with greater precision using diffraction software packages.

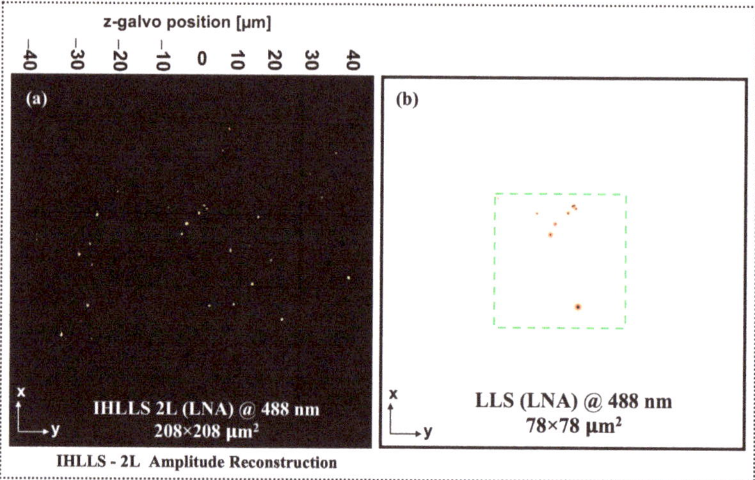

Figure 6. Volume imaging comparison between the IHLLS 2L and LLS with LNA; (**a**) IHLLS 2L bead volume reconstruction; (**b**) LLS bead volume imaging. The green rectangle in the middle of the image represents the scanning area of the conventional LLS system in the dithering mode.

4. Conclusions

The approach of using lower-NA (LNA) masks to generate lattice light sheets forms beams that approximate Gaussian rather than Bessel beams. This provides a tradeoff between the size of the imaged beam and the resolution of the ultimate image in the z-plane. This has advantages for imaging objects at larger intercellular scales. Here, we demonstrate that even with these longer but lower-resolution beams, the IHLLS approach to generate holograms to resolve 3D positional information is functional. In fields of study such as neuroscience, where the cell structure is complex but at large scales, this approach will enable the resolution of complex 3D structures capturing structures as large as 200 μm in length.

Author Contributions: Conceptualization, M.P. and S.A.; methodology, M.P. and S.A.; software, M.P. and C.M.; validation, M.P., S.A., J.A. and C.M.; resources, S.A., J.A. and C.M.; data curation, M.P., S.A. and C.M.; writing—original draft preparation, M.P. and S.A.; writing—review and editing, M.P., S.A., J.A. and C.M.; visualization, M.P., S.A., J.A. and C.M.; supervision, S.A. and J.A.; project administration, S.A. and J.A.; funding acquisition, S.A. and J.A. All authors have read and agreed to the published version of the manuscript.

Funding: This research was funded by the NIH RO1 NS111749 to S.A. and the NIH R21 DC017292 to J.A.

Institutional Review Board Statement: All work on animals was performed according to Institutionalization guidelines in an AALAC-approved facility at UIC. The work was approved by the institutional IACUC review committee under protocol number ACC 20-119 in January 2022.

Informed Consent Statement: Not applicable.

Data Availability Statement: Supporting data for Figures 4–6 are available from the corresponding author on reasonable request. This is due to the size of the datasets being so large that they are not available on a public server.

Acknowledgments: We would like to thank Lisa Hoffman for her support in the laboratory.

Conflicts of Interest: The authors declare no conflict of interest.

References

1. Huisken, J.; Swoger, J.; Del Bene, F.; Wittbrodt, J.; Stelzer, E.H.K. Optical Sectioning Deep Inside Live Embryos by Selective Plane Illumination Microscopy. *Science* **2004**, *305*, 1007–1009. [CrossRef]
2. Davidovits, P.; Egger, M.D. Scanning Laser Microscope. *Nature* **1969**, *223*, 831. [CrossRef] [PubMed]
3. Pampaloni, F.; Ansari, N.; Girard, P.; Stelzer, E.H.K. Light sheet-based fluorescence microscopy (LSFM) reduces phototoxic effects and provides new means for the modern life sciences. In Proceedings of the Advanced Microscopy Techniques II, Munich, Germany, 22 May 2011; p. 80860Y.
4. Laissue, P.P.; Alghamdi, R.A.; Tomancak, P.; Reynaud, E.G.; Shroff, H. Assessing phototoxicity in live fluorescence imaging. *Nat. Methods* **2017**, *14*, 657–661. [CrossRef] [PubMed]
5. Stelzer, E.H.K.; Strobl, F.; Chang, B.-J.; Preusser, F.; Preibisch, S.; McDole, K.; Fiolka, R. Light sheet fluorescence microscopy. *Nat. Rev. Methods Prim.* **2021**, *1*, 73. [CrossRef]
6. Gao, L.; Shao, L.; Chen, B.-C.; Betzig, E. 3D live fluorescence imaging of cellular dynamics using Bessel beam plane illumination microscopy. *Nat. Protoc.* **2014**, *9*, 1083–1101. [CrossRef] [PubMed]
7. Gao, L. Optimization of the excitation light sheet in selective plane illumination microscopy. *Biomed. Opt. Express* **2015**, *6*, 881–890. [CrossRef] [PubMed]
8. Chen, B.-C.; Legant, W.R.; Wang, K.; Shao, L.; Milkie, D.E.; Davidson, M.W.; Janetopoulos, C.; Wu, X.S.; Hammer, J.A.; Liu, Z.; et al. Lattice light-sheet microscopy: Imaging molecules to embryos at high spatiotemporal resolution. *Science* **2014**, *346*, 1257998. [CrossRef] [PubMed]
9. Potcoava, M.; Mann, C.; Art, J.; Alford, S. Spatio-temporal performance in an incoherent holography lattice light-sheet microscope (IHLLS). *Opt. Express* **2021**, *29*, 23888–23901. [CrossRef] [PubMed]

Proceeding Paper

Imaging Incoherent Target Using Hadamard Basis Patterns †

Tanushree Karmakar *, Rajeev Singh and Rakesh Kumar Singh

Laboratory of Information Photonics and Optical Metrology, Department of Physics,
Indian Institute of Technology (Banaras Hindu University), Varanasi 221005, India;
rajeevs.phy@itbhu.ac.in (R.S.); krakeshsingh.phy@iitbhu.ac.in (R.K.S.)
* Correspondence: tanushreekarmakar.rs.phy20@itbhu.ac.in
† Presented at the International Conference on "Holography Meets Advanced Manufacturing", Online, 20–22 February 2023.

Abstract: In this paper, we present a correlation-based imaging technique in a single-pixel imaging scheme using Hadamard basis illumination. The Hadamard basis, which has the characteristics of a two-bit value $\{-1, 1\}$ and sparsity in its transformed domain, has been used in the illumination patterns and successfully utilized for imaging the incoherent target. It gives image reconstruction even in low-light conditions. Such deterministic patterns also help to solve the problem of large numbers of measurements in single-pixel imaging, and hence simplify the experimental implementation. Furthermore, to compare the quality of imaging with Hadamard basis patterns, we also compare imaging with Fourier basis patterns and simulation results of both methods, namely Hadamard and Fourier basis, are presented.

Keywords: single-pixel imaging; Hadamard basis patterns; Fourier basis patterns

Citation: Karmakar, T.; Singh, R.; Singh, R.K. Imaging Incoherent Target Using Hadamard Basis Patterns. *Eng. Proc.* **2023**, *34*, 17. https://doi.org/10.3390/HMAM2-14271

Academic Editor: Vijayakumar Anand

Published: 31 March 2023

1. Introduction

The correlation-based imaging technique is an unconventional approach where an object is reconstructed as the distribution of correlations of the light field to image the desired target [1]. Single-pixel imaging, part of the unconventional techniques, is a novel imaging scheme where an object can be imaged by a single-pixel detector without having any spatial resolution. Pixelated detectors are mostly expensive and not capable of detecting signals on the spectrum, such as x-ray and infrared. Moreover, a low-cost imaging system can be designed with a single-pixel detector that can operate at other wavelength regions, even in low-light conditions, and techniques such as ghost imaging, and non-line of sight (NLOS) imaging have greatly benefited from single-pixel detection [2].

Ghost imaging (GI) provides object information from correlation-based measurements between two light beams. One light beam that interacts with the object to be imaged is collected by a bucket detector having no spatial resolution. Another beam is collected with a spatially resolved detector, such as a charged coupled device (CCD), which samples the light that has never interacted with the object by an array of pixels. The GI was first demonstrated using the correlation of quantum entangled photons generated by spontaneous parametric downconversion. It was later demonstrated that GI can be performed with a classical incoherent (pseudothermal) light source in the same way as quantum entangled photons. Computational ghost imaging is very similar to single-pixel imaging, where an object is illuminated with digitally structured light patterns, and corresponding to each light pattern, the single-pixel detector records light signals [3]. The illumination light patterns are projected onto the object by a spatial light modulator (SLM) or digital micro-mirror device (DMD). The light patterns can be random or can be deterministic illumination patterns that can reduce the measurement number and acquisition time in a single-pixel imaging scheme. The Hadamard patterns are such patterns; they are made of orthogonal basis vectors. The Hadamard matrix has the characteristics of a 2-bit binary value $\{-1, 1\}$ and sparsity in its transformed domain. In Hadamard single-pixel imaging (HSI), Hadamard matrices form illumination patterns to reconstruct the image of an object [4]. Fourier single-pixel imaging (FSI) is also another robust imaging technique based on a deterministic model that uses Fourier basis patterns [5].

In this work, we present a single-pixel correlation imaging technique using Hadamard illumination to image an incoherent target object. Such deterministic illuminations help in reducing the number of measurements and acquisition time. Object retrieval in the FSI with the four phase shifts requires $4 \times M \times M$ number of measurements, which can be reduced to the $2 \times M \times M$ number of measurements for the imaging with the Hadamard pattern by a differential measurement approach. $M \times M$ is the size of the image that has to be reconstructed. Hadamard basis patterns are binary (black and white), which makes HSI naturally suitable for single-pixel imaging systems. HSI is more noise-robust compared to FSI [2]. In order to qualitatively compare imaging with Hadamard and the Fourier basis patterns, we present simulation results for both cases. This reconstruction highlights that the Hadamard illumination patterns can be an alternative to FSI depending on the nature of requirements.

2. Theory

In the HSI scheme, the Hadamard transform $(H\{\ \})$ of an object $A(x, y)$ is expressed as [6]

$$H\{A(x,y)\} = \sum_{x=0}^{L-1}\sum_{y=0}^{L-1} A(x,y)(-1)^{w(x,y,u,v)} \tag{1}$$

where x,y are coordinates in spatial domain and u,v are coordinates in Hadamard domain. L is the order of the Hadamard matrix and L is number of rows and columns of the image.

$$w(x,y,u,v) = \sum_{i=0}^{n-1}[g_i(u)x_i + g_i(v)y_i] \tag{2}$$

$$\begin{aligned} g_0(u) &= u_{n-1} \\ g_1(u) &= u_{n-1} + u_{n-2} \\ g_2(u) &= u_{n-2} + u_{n-3} \\ &\ \cdots \cdots \\ g_{n-1}(u) &= u_1 + u_o \end{aligned} \tag{3}$$

where $n = \log_2 N$ and $u_i,\ v_i,\ x_i,\ y_i$ are the binary representations of u, v, x, and y, respectively. Hadamard basis pattern $P_H(x, y)$ is presented as follows

$$P_H(x,y) = \frac{1}{2}\left[1 + H^{-1}\{\delta_H(u,v)\}\right] \tag{4}$$

$$\delta_H(u,v) = \begin{cases} 1, if\ u = u_0\ and\ v = v_0 \\ 0, otherwise \end{cases}$$

where H^{-1} represents inverse Hadamard transform and $\delta_H(u,v)$ is the delta function.

In order to consider the imaging of an incoherent target, we consider an object obscured by a rotating diffuser, as shown in Figure 1. The incoherent target is realized by a sequence of a random scatterer, which corresponds to each step of the rotating diffuser. A step of the diffuser is considered to introduce an independent and a uniformly distributed phase on the interval $(-\pi,\pi)$. An object is illuminated by the Hadamard patterns $P_H(x,y)$, and a single-pixel detector (SPD) measures the responses corresponding to each pattern. Similarly, inverse Hadamard patterns, i.e., $1 - P_H(x,y)$, are also projected in sequence onto the object, and the same process is followed to record the responses at the SPD. D_+ and D_- are the SPD responses corresponding to the patterns $P_H(x,y)$ and $1 - P_H(x,y)$. Then, the Hadamard coefficient is derived from the two responses, i.e., D_+ and D_-, by this differential measurement formula.

$$H(u,v) = D_+ - D_- \tag{5}$$

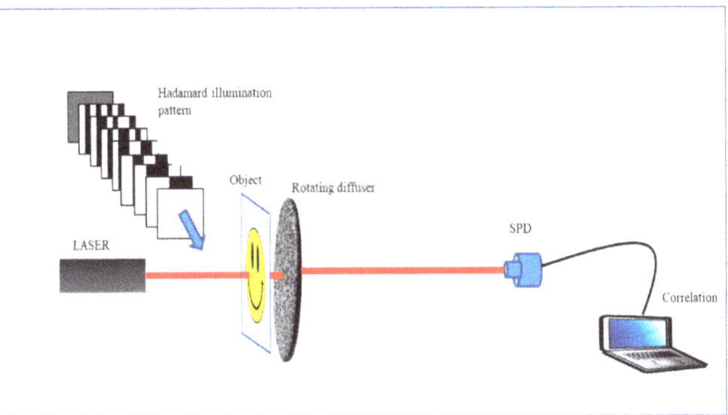

Figure 1. Schematic diagram of Hadamard single-pixel imaging of incoherent target object.

The object is reconstructed by taking the inverse Hadamard transformation of $H(u, v)$. A schematic diagram of the Hadamard illumination with the SPD is shown in Figure 1, where an object is placed behind a rotating diffuser. Hadamard patterns and inverse Hadamard patterns illuminate the object, as shown in Figure 1. SPD collects signals corresponding to each pattern. After correlation of these patterns, objects are reconstructed.

3. Results and Discussion

We have shown here image reconstruction for an incoherent object "1" using both HSI and FSI in Figure 2. Simulation results using MATLAB are presented below. To image the incoherent target of size 256×256, $2 \times 256 \times 256$, the number of Hadamard patterns, including Hadamard patterns ($P_H(x, y)$) and inverse Hadamard patterns ($1 - P_H(x, y)$), are projected onto the object. Hadamard patterns are formed according to equation number four. Similarly, Fourier-structured light is projected onto the object of size 256×256. To reconstruct the object, a total $4 \times 256 \times 256$ number of Fourier patterns is needed to illuminate the object to apply four phase shifts in the FSI scheme.

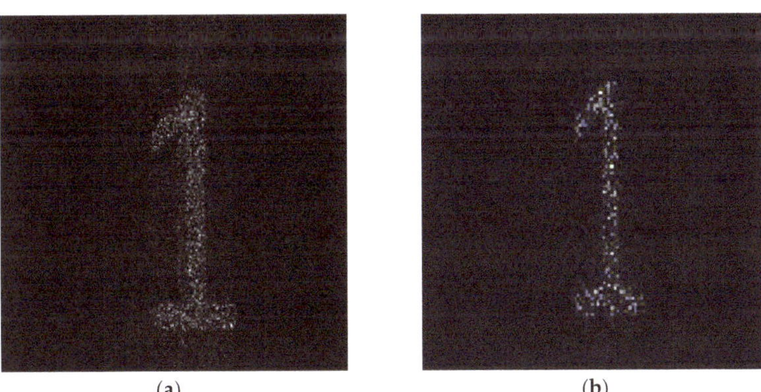

(a) (b)

Figure 2. Reconstruction of an incoherent object using (a) HSI (b) FSI.

4. Conclusions

In summary, a correlation-based single-pixel imaging technique using Hadamard illumination is presented. The Hadamard patterns are formed with an orthogonal basis vector consisting of two binary values, +1 and −1. Such a deterministic pattern improves the single-pixel imaging scheme by solving the problem of having a large number of measurements. We have successfully used Hadamard basis illumination to image an incoherent target hidden behind the diffuser. A differential measurement of Hadamard coefficient reduces noise, and it is easier than phase shifting in the case of the Fourier basis pattern. Furthermore, to compare reconstruction qualities, we have also shown simulation results using the Fourier basis pattern.

Author Contributions: T.K.: Conceived the idea, investigation, simulation, data analysis, writing manuscript; R.S.: Editing and revision of manuscript; R.K.S.: Ideas, formulation of research goals, revision and editing, funding acquisition, supervision. All authors have read and agreed to the published version of the manuscript.

Funding: This work is supported by the Science and Engineering Research Board (SERB) India CORE/2019/000026.

Institutional Review Board Statement: Not applicable.

Informed Consent Statement: Not applicable.

Data Availability Statement: The data used in this paper are not currently publicly accessible, although they are available from the authors upon reasonable request.

Acknowledgments: Tanushree Karmakar would like to acknowledge support from DST-INSPIRE (IF190840).

Conflicts of Interest: The authors declare no conflict of interest.

References

1. Takeda, M.; Wang, W.; Duan, Z.; Miyamoto, Y. Coherence holography. *Opt. Express* **2005**, *13*, 9629–9635. [CrossRef] [PubMed]
2. Zhang, Z.; Wang, X.; Zheng, G.; Zhong, J. Hadamard single-pixel imaging versus Fourier single-pixel imaging. *Opt. Express* **2017**, *25*, 19619–19639. [CrossRef] [PubMed]
3. Yang, X.; Liu, Y.; Mou, X.; Hu, T.; Yuan, F.; Cheng, E. Imaging in turbid water based on a Hadamard single-pixel imaging system. *Opt. Express* **2021**, *29*, 12010–12023. [CrossRef] [PubMed]
4. Singh, R.K. Hybrid correlation holography with a single pixel detector. *Opt. Lett.* **2017**, *42*, 2515–2518. [CrossRef] [PubMed]
5. Mandal, A.C.; Sarkar, T.; Zalevsky, Z.; Singh, R.K. Structured transmittance illumination coherence holography. *Sci. Rep.* **2022**, *12*, 4564. [CrossRef] [PubMed]
6. Pratt, W.K.; Kane, J.; Andrews, H.C. Hadamard transform image coding. *Proc. IEEE* **1969**, *57*, 58–68. [CrossRef]

Proceeding Paper

Design and Simulation of a Low-Power and High-Speed Fast Fourier Transform for Medical Image Compression [†]

Ernest Ravindran Ramaswami Sachidanandan *, Ngangbam Phalguni Singh and Sudhakiran Gunda

Department of Electronics and Communication Engineering, Koneru Lakshmaiah Education Foundation, Vaddeswaram, Guntur 522302, India; npsingh@kluniversity.in (N.P.S.); sudhakiran.495@gmail.com (S.G.)
* Correspondence: ravindran.ernest@kluniversity.in
† Presented at the International Conference on "Holography Meets Advanced Manufacturing", Online, 20–22 February 2023.

Abstract: For front-end wireless applications in small battery-powered devices, discrete Fourier transform (DFT) is a critical processing method for discrete time signals. Advanced radix structures are created to reduce the impact of transistor malfunction. To develop DFT, with radix sizes 4, 8, etc., is a complex and tricky issue for algorithm designers. The main reason for this is that the butterfly algorithm's lower-radix-level equations were manually estimated. This requires the selection of a new design process. As a result of fewer calculations and smaller memory requirements for computationally intensive scientific applications, this research focuses on the radix-4 fast Fourier-transform (FFT) technique. A new 64-point DFT method based on radix-4 FFT and multi-stage strategy to solve DFT-related issues is presented in this paper. Based on the results of simulations with Xilinx ISE, it can be concluded that the algorithm developed is faster than conventional approaches, with an 18.963 ns delay and 12.68 mW of power consumption. It was found that the computed picture compression drop ratios of 0.10, 0.31, 0.61 and 0.83 had a direct relationship to the varied tolerances tested, 0.0007625, 0.003246, 0.013075 and 0.03924. Fast reconstruction techniques, wireless medical devices and other applications benefit from this FFT's low power consumption, small storage requirements, and high processing speed.

Keywords: discrete Fourier transform; inverse discrete Fourier transform; fast Fourier transform; inverse fast Fourier transform; radix-4; image compression; drop ratio

Citation: Ramaswami Sachidanandan, E.R.; Phalguni Singh, N.; Gunda, S. Design and Simulation of a Low-Power and High-Speed Fast Fourier Transform for Medical Image Compression. *Eng. Proc.* **2023**, *34*, 18. https://doi.org/10.3390/HMAM2-14159

Academic Editor: Vijayakumar Anand

Published: 13 March 2023

1. Introduction

One of the most widely utilized mathematical operations is fast Fourier transform. Several medical applications use fast Fourier transform for image reconstruction and frequency domain analysis. Image processing applications, such as filtering, compression, and de-noising, all rely on FFT to a certain extent. FFT, is used as an improved version of the traditional discrete signal processing tool (discrete Fourier transform), for medical image compression with various drop ratios. FFT is widely used in medical imaging, engineering, communication, and other fields because it transitions quickly from the T-domain to the F-domain and vice versa [1–6].

The medical imaging method provides images of the human body and its components for clinical application. Computer tomography (CT), magnetic resonance imaging (MRI), ultrasound and optical imaging technology are the most prevalent modes of medical imaging that produce a prohibitive amount of data. The images produced by these instruments are pixels representing the operations of human organs in terms of their visual depiction. They are also the patient's most vital information and demand high storage and transmission width [7,8].

FFT-based compression is a compression algorithm, which can process the image quickly coupled with the transformed domain compression. The modified domain includes coefficients of both low and high frequencies that are measured. Various quantified coefficient values of high frequency are unimportant and almost equivalent to zero and remove them from the modified image. This pre-processing step leads to the compression platform. By supplying different symbols, the FFT method accomplishes compression. The majority of appearing symbols are assigned to be shorter while the rest are assigned to longer-size symbols. The variable-length compressed data are subsequently stored on transmission media.

As hospitals are progressing into digitization, filmless imaging and telemedicine, medical imagery becomes significantly important in the health sector. This has led to the major difficulty of developing compression algorithms that prevent diagnostic errors and have a high compression ratio for lower bandwidth and storage. In the medical area, particularly, quick diagnosis is only achievable when the required diagnostic information is maintained through the compression approach. These images help physicians to easily diagnose the inner parts of the body. They also help to perform keyhole procedures without too many incisions to reach the inner sections of the body. They can be processed quickly, analyzed objectively and made available in numerous places simultaneously by means of communication protocols and networks, such as digital imaging and communications in medicine (DICOM) protocol and picture archiving and communication system (PACS) networks, respectively. The X-ray, CT, MRI or ultrasound images contain huge amounts of data that demand vast channels or storage capacities. The implementation costs limit the storage capacity, even with the progress in storage capacity and connectivity [8–10]. There are certain approaches that create imperceptible variations and acceptable fidelity that can lead to medical picture low-loss compression. In this article, FFT-based compression is proposed.

Many different mathematical FFT algorithms vary from easy theories of complex numbers arithmetically to group and numerical theory; this paper provides an available technical outline and few characteristics while explaining the algorithms in the subsidiary sections. The DFT is obtained via the decomposition of a series of values into various frequency components, as given in Equations (1) and (2). This operation is useful in several fields, but it is always too slow to be practical for computing it directly from the given description.

$$X(k) = \sum_{n=0}^{N-1} x(n)e^{-j(\frac{2\pi}{N})kn}, \ 0 \leq k \leq N-1 \tag{1}$$

$$x(n) = 1/N \sum_{k=0}^{N-1} X(k)e^{j(\frac{2\pi}{N})kn}; \ 0 \leq n \leq N-1 \tag{2}$$

The FFT is one of the new ways to calculate similar results faster: DFT takes N^2 arithmetical multiplications and $N^2 - N$ addition operations ($O(4N^2)$, real multiplications and $O(N(4N-2)$ real additions) to naively compute the DFT of N-points. The speed difference may be huge, especially in long data sets where N may be higher. FFT can compute the DFT for $\frac{N}{2}\log_r N$ multiplications and $N\log_r N$ additions corresponding operation alone using twiddle factor $W_N = e^{-j2\pi/N}$ [4,6] since it is using a butterfly operation and computes $p \pm \alpha q$ (results six real adds and four real multiplications). As FFTs are staged algorithms, there are log_2^N stages, and each stage has $N/2$ butterflies, so there should be four.

$$N/2 \times log_2^N = 2N \times log_2^N \text{ real multiplications and } 3N \times log_2^N \text{ realadditions for N-point DFT through FFT} \tag{3a}$$

(Optimization is still possible, but these are basic equations).

$$N^2 \text{ and } N(N-1) \text{ multiplications and additions in a normal DFT for N-point} \tag{3b}$$

DFT estimation was practical due to these huge changes. For a broad range of applications, FFTs are of great importance—from DSP to algorithms for quick multiplication of high integer range [10,11]. From Equations (3a) and (3b), the cost estimation is given in Table 1.

Table 1. Cost estimation of DFT and FFT.

N	DFT		FFT (Radix 2) (Optimized)		Speed Factor Improvement	
	$4N^2$	$N(4N-2)$	$2Nlog_2^N$	$3Nlog_2^N$	Multiplications	Additions
2	16	12	4	6	4	2
32	4096	4032	320	480	12.8	8.3
64	16,384	16,256	768	1152	21.333	14.11
1024	41,94,304	41,92,256	20,480	30,720	204.8	136.466

From Table 1, it is evident that FFT is less computationally intensive than DFT. When comparing the two methods, FFT is faster. It is important to note that, because FFTs are radix algorithms, this work shows that making minor changes to the algorithm in the order 2 results in faster FFT times. A DIT-FFT algorithm decomposes a signal based on the time sequence '$x(n)$'. Another categorization is the decimation-in-frequency FFT (DIF-FFT) algorithm, which decomposes using the frequency sequence '$X(k)$'. Radices are the foundation of these algorithms. Many intermediate results and memory locations are re-used in these algorithms, which makes them more efficient in the long run. These computational approaches happen with the help of butterflies called radix-2 butterflies.

In DIT-FFT, the inputs are arranged in bit reversal/normal order, and outputs are obtained in a normal order/bit reversal. The radix-2 DIT-FFT algorithm is a staged algorithm. The effective functioning of radix-2 depends on stages, butterflies, etc. Each stage has block(s), and each block has butterflies. These can be defined as follows, radix-2 algorithm consists of log_2N stages, and each stage consists of $N/2^{stage}$ blocks, and each block consists of $2^{stage-1}$ butterflies. As in signal processing (digital), most of the required arithmetic computations are additions and multiplications, radix-2 offers $N/2 \times log_2 N$ complex multiplications and $N \times log_2 N$ complex additions [9–14].

The radix-4 is an additional fast Fourier transform algorithm (FFT) that can be obtained by moving the base from 2 to 4. The power/index diminishes in direct proportion to the size of the base. There are 50% fewer stages in radix-4 than in radix-2 since $N = 4M$, indicating that stages have decreased by 50%. It is explained in more detail in the later sections on how radix-4 simplifies difficult calculations [15]

For computing sequences, the radix-4 algorithm is comparable to the radix-2 technique in terms of type and speed management. It takes place as follows: the given sequence divides into four parts based on 'n': The given sequence layout in radix-4 is as follows:

n = [0, 4, 8, ... N − 4] results × (4n),
n = [1, 5, 9, ... N − 3] results × (4n + 1),
n = [2, 6, 10, ... N − 2] results × (4n + 2),
n = [3, 7, 11, ... N − 1] results × (4n + 3) [16–18].

After the division of N-point DFT, it can be computed as the sum of the outputs of four N/4-point DFTs, and these sub-sequences are interconnected with so-called twiddle factors $W^{lk}{}_N = e^{-(j2l\pi k/N)}$, $l = 0, 1, 2, 3$, as shown in (4).

$$X(k) = \sum_{n=0}^{N-1} x(n)W_N^{nk} = \sum_{n=0}^{\frac{N}{4}-1} x(n)W_N^{nk} + \sum_{n=\frac{N}{4}}^{\frac{N}{2}-1} x(n)W_N^{nk} + \sum_{n=\frac{N}{2}}^{\frac{3N}{4}-1} x(n)W_N^{nk} + \sum_{n=\frac{3N}{4}-1}^{N-1} x(n)W_N^{nk} \tag{4}$$

Thus,

$$X(k) = \sum_{n=0}^{\frac{N}{4}-1} x(n)W_N^{nk} + W_N^{\frac{Nk}{4}}\sum_{n=\frac{N}{4}}^{\frac{N}{2}-1} x\left(n+\frac{N}{4}\right)W_N^{nk} + W_N^{\frac{Nk}{2}}\sum_{n=\frac{N}{2}}^{\frac{3N}{4}-1} x\left(n+\frac{N}{2}\right)W_N^{nk} + W_N^{\frac{3Nk}{4}}\sum_{n=\frac{3N}{4}-1}^{N-1} x(n)W_N^{nk} \tag{5}$$

Final representation of X(k) is,

$$X(k) = \sum_{n=0}^{\frac{N}{4}-1} \left[x(n) + (-j)^k x\left(n + \frac{N}{4}\right) + (-1)^k x\left(n + \frac{N}{2}\right) + (j)^k x(n + \frac{3N}{4}) \right] W_N^{nk} \quad (6)$$

According to Equations (4)–(6), the process is called decimation in time because the samples of time are arranged into groups. The basic operation of the R4 butterfly is shown in Figure 1 [17]. The decimation-in-time process consolidates the inputs at each stage of decomposition, resulting in "input order that is bit-reversed" at the end. This set-up allows for the intermediate outputs to be stored in the same memory regions as the inputs (in-place algorithm). The radix-4 FFT's slight reorganization allows the inputs to be redirected from digit to bit [18], as shown in Table 2.

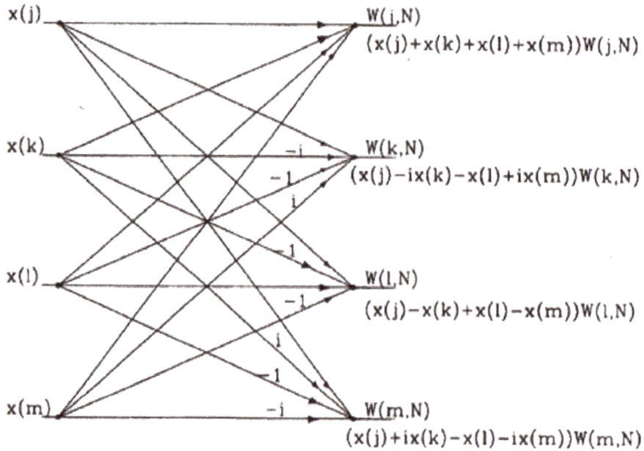

Figure 1. Operation of radix-4 butterfly (radix 4 nodes—dragon nodes).

Table 2. Numbering according to base-4 for bit reverse.

Normal Sequence Order	Normal Sequence Bit Order	Reversed Bit Order	Reversed Sequence Number
0	000	000	0
1	001	100	16
2	002	200	32
3	003	300	48
4	010	010	4
5	011	110	20
14	032	230	44

Figure 2 depicts the computation of the radix-4 project's flow chart. The input sequence might be in bit reversal order or normal order. The updated sequence can be operated on after being arranged. Each stage has a group of butterflies, and each butterfly group is made up of other butterflies. After that, the butterfly (radix-4) algorithm is used. For each further stage or group, the operation repeats until all butterflies in a group and stages have been completed by scaling with the required twiddle factor. Here, the radix-4 operation is completed with the output in either a normal or bit-reversed order depending on the input sequence [15–18] and radix-4 operations are completed.

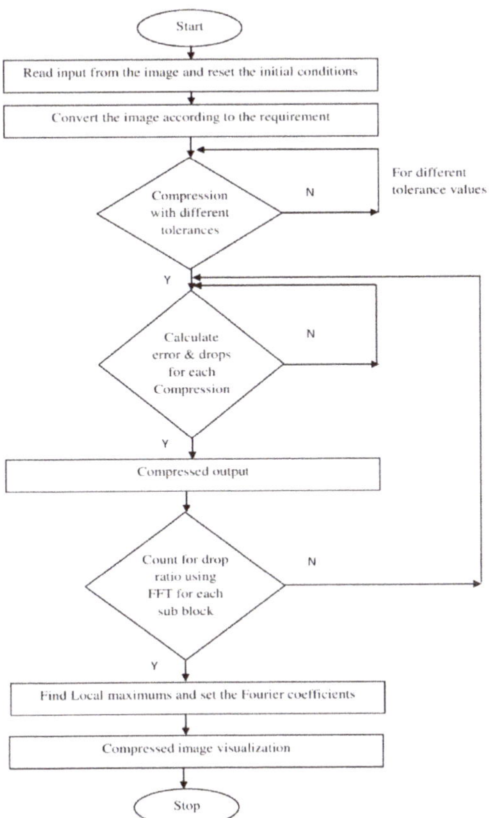

Figure 2. Medical image compression using FFT algorithm flowchart.

The radix-2 adds twiddle integer factors at $0°$ and $180°$ angles, whereas radi-x4 adds twiddle integer factors at $0°$, $90°$, $180°$ and $270°$ angles, all while accounting for the computational cost of multiplication. There is no need to multiply the sine and cosine counterparts of the above-mentioned angles within a unit circle. The radix-8 is not preferred because of its factors of fractional twiddle 2 at $45°$, $135°$, $225°$ and $315°$ in a unit circle, despite the fact that the number of radixes minimizes the number of computation steps [19,20].

2. Development of a Lossless Medical Image Compression Using FFT Algorithm

Medical image compression using FFT is developed as shown in the flow chart given in Figure 3. Load/read any medical input image and convert it into a 2D array of a double image. Compress the loaded image with different tolerance values. While compressing the image, it takes as inputs the original image X and the drop tolerance parameter and outputs a compressed image Y. It also returns the drop ratio given in Equation (7), which is defined to be as the ratio of "Total number of nonzero Fourier coefficients dropped to the Total number of initially nonzero Fourier coefficients".

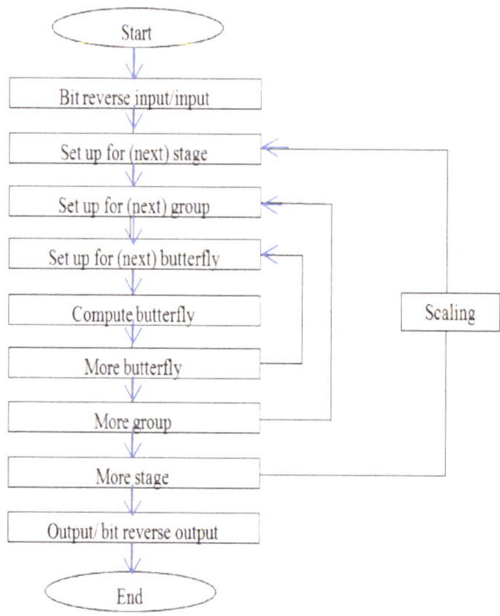

Figure 3. Flow chart of radix-4 64-point DIT-FFT operation.

For every drop count for a compressed image, apply FFT to each sub block.

$$\text{Drop ratio} = (\text{Total number of nonzero Fourier coefficients dropped}/\text{Total number of initially nonzero Fourier coefficients}) \tag{7}$$

3. Results and Discussion

A 64-Point DFT using the radix-4 DIT-FFT algorithm has three stages. In the first stage, 16 blocks are present, and each block consists of only one butterfly. In the first stage, the inputs are applied in a bit-reversal order to save memory space. In the second stage, there are four blocks, and each block consists of four butterflies as a set. Totally, four sets are present. In the third stage and in the final stage, only one block is present in that 16 butterflies are present as one set, and the output is finally obtained from the final stage, which is in a normal order. A structural view of the 64-point radix-4 DIT-FFT is shown in Figure 4. The Register-Transfer Level (RTL) view of the proposed algorithm consists of data splitting of 256 bits into 4 64 bits, and each section of 64 bits is further divided into 16 4-bit points and separated into even and odd sequences. All are communicated with a communicator called Commutator.

The target XC3S500E-5FGG320C Xilinx FPGA, Guntur, India is used for the execution. The device contains the 9312 LUTs and 4656 slices for functionality of the input sequence. Slices used for the related logic are 3286 and 3286 for the unrelated logic. The device contains 9312 four-input LUTs and works under a speed grade of '-5'.

Generally, the 64-point radix-4 DIT-FFT butterfly unit (butterfly 16) consists of butterflies, sets of butterflies and stages (Figure 4). The entire code was developed in a structure made using HDL language. The internal modules were designed based on the behavioral model or dataflow model. The different internal modules are: splitting the entire sequence into equal parts to save memory, 4-bit butterfly unit, odd and even parts, butter for 8-point and 4-point and Commutator for connecting all the stages and sub modules. All these modules are called sub modules in the top butter. The RTL view and simulation result of the 64-point radix-4 DIT-FFT are shown in Figure 5.

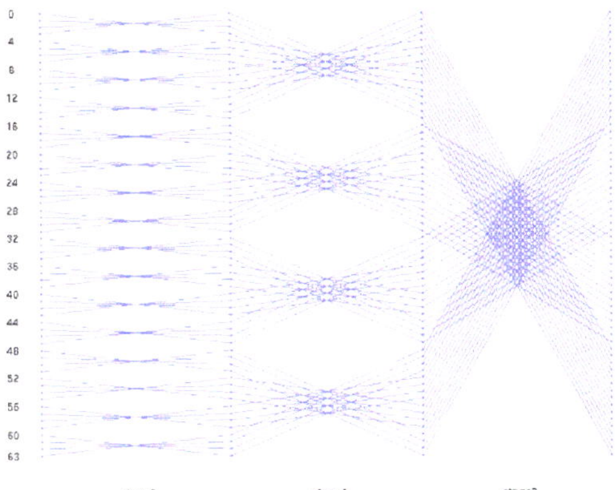

Figure 4. Butterfly diagram of 64-point DFT using radix-4 DITFFT.

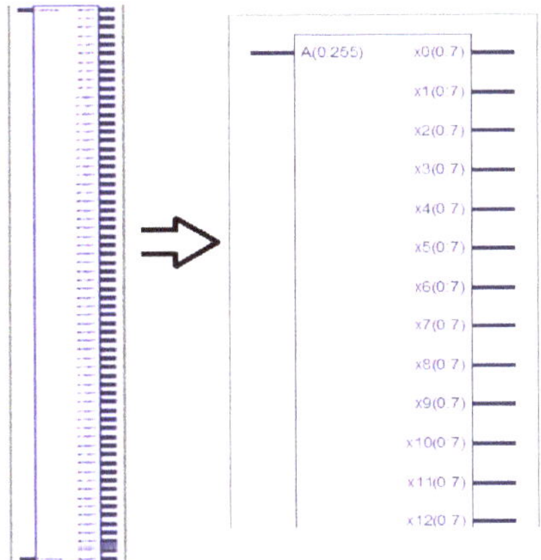

Figure 5. RTL view of 256-bit 64-point radix-4 DITFFT.

3.1. Data Split of 64 Points

The entire 64-point radix-4 DIT-FFT can be split into four equal parts. Each of the 16 points and each point are the combination of 4 bits; totally, 64 bits are in each equal part. A is an input sequence of 256 bits (0 to 255), divided into four equal parts of each 16 point 64 bit (0 to 63, 64 to 127, 128 to 191, 192 to 255). The RTL view of data split 64 points into four equal 16 points, and the simulation results are shown in Figures 6–8, respectively.

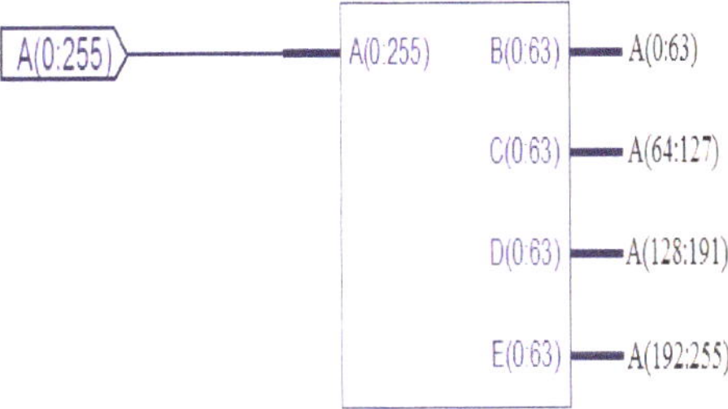

Figure 6. RTL view of data split of 64 points.

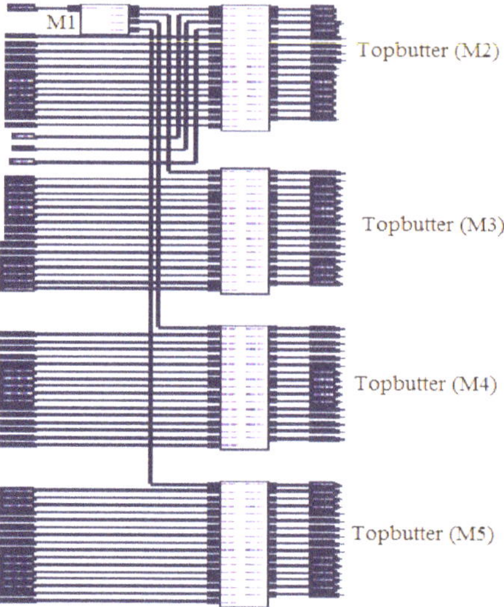

Figure 7. Internal structure of 64-point radix-4 DIT-FFT.

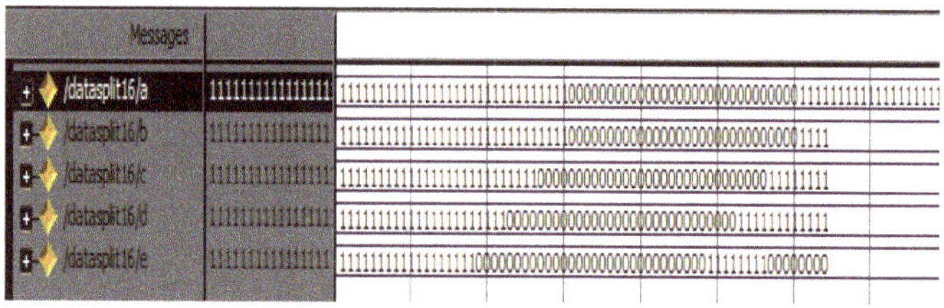

Figure 8. Simulation results of data split of 64 points.

3.1.1. Top Butter

Top butter is the divided part of the entire sequence. This has a 16-point, 64-bit input and 64 twiddle, totally 128-bit output, in which 64 bits are even and the remaining bits are odd. Top butter contains the Commutator, even and odd parts. For the entire sequence, the total 4-top butters (M2, M3, M4 and M5) are present. The RTL view of the top butter (M2), and the simulation results are shown in Figures 9 and 10, respectively:

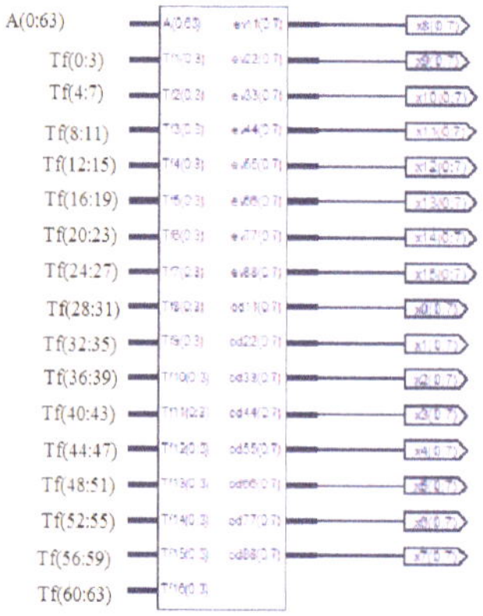

Figure 9. Overview of top butter (M2).

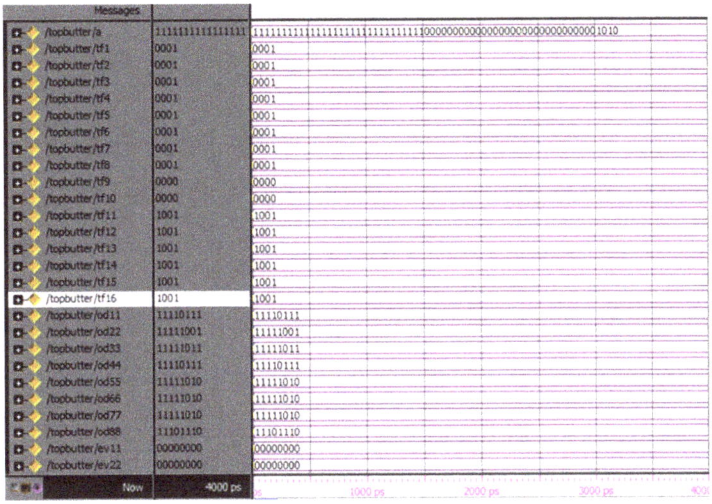

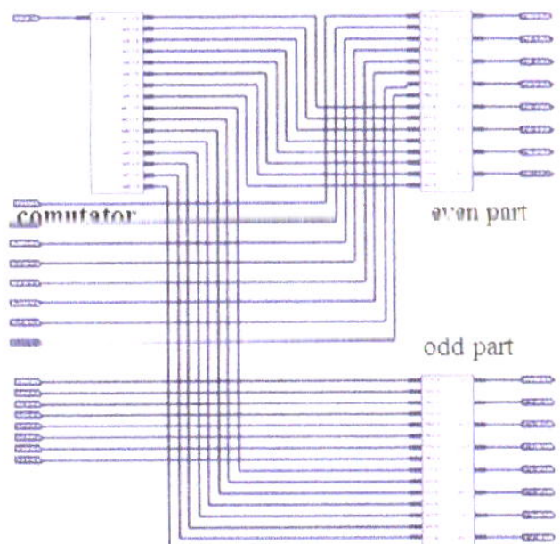

Figure 10. Simulation results of top butter odd and even part.

3.1.2. Internal Structure of Top Butter

The equally divided 16-point (64-bit) sequence again undergoes further division to increase the speed of execution. The 64-bit (0 to 63) sequence divides as A (0 to 3), A (4 to 7) and so on to A (59 to 63), and 16 points are divided into the remaining four equal parts. The simulation results of data split of 16 points are shown in Figures 11 and 12 and RTL view of this unit having Commutator (Figures 13 and 14), even and odd parts respectively:

Figure 11. Internal view of top butter (M2).

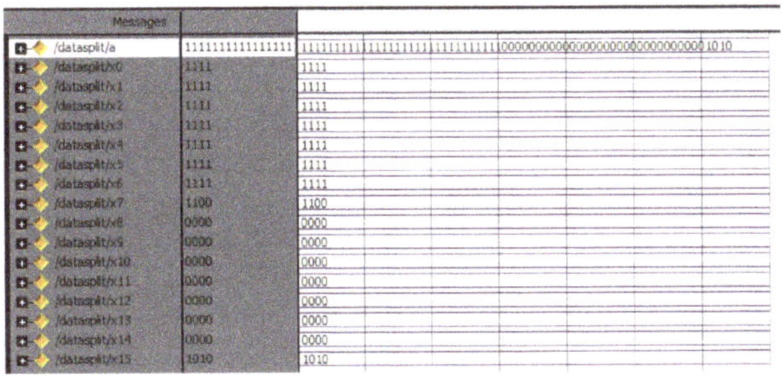

Figure 12. Simulation results of top butter data split.

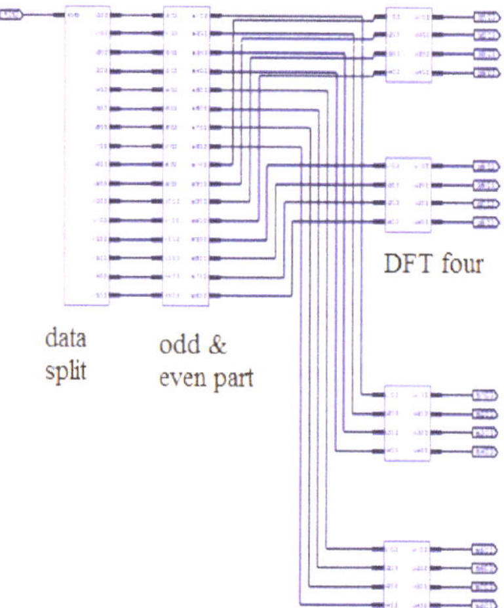

Figure 13. Internal structure of commutator.

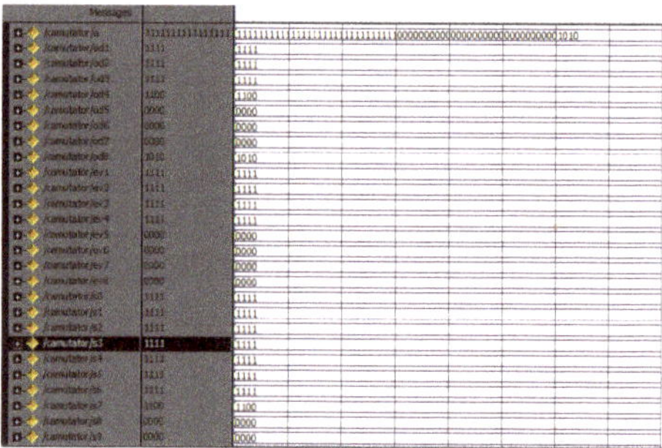

Figure 14. Simulation results of Commutator.

3.1.3. Commutator

As commutator is used to communicate, it consist of data splitter, odd and even parts, DFT four bit odd and even parts separately as shown in Figure 13 and its simulated results is shown in Figure 14.

3.1.4. DFT Four

DFT four is the basic unit in the radix-4 structure, because it transfers the input value to output. Each stage has this unit, and four inputs and four outputs are present in this unit. The RTL view and simulation results of DFT four are shown in Figures 15 and 16, respectively.

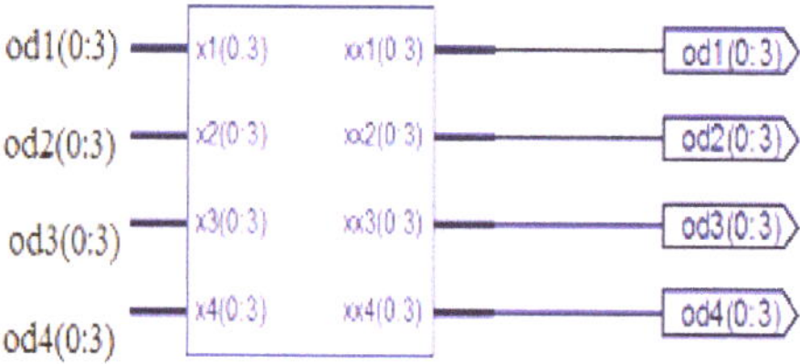

Figure 15. DFT four block.

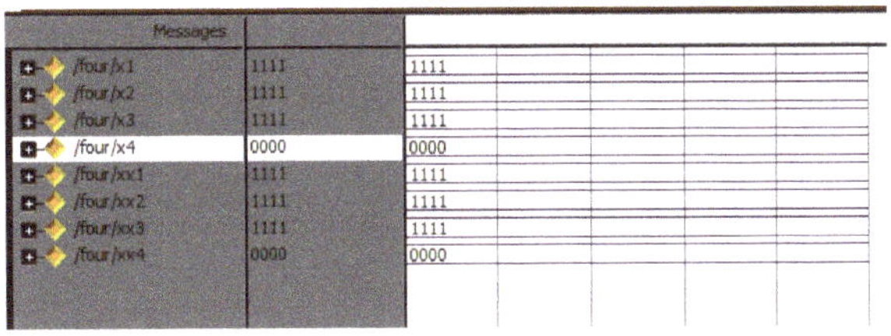

Figure 16. Simulation results of DFT four.

3.1.5. Butter R8

Butter R8 is the internal unit in the butterfly diagram. It is the combination of both even and odd parts. Each even and odd part is the combination of two butter R4 blocks, so there are a total of two R4 blocks for each butter R8 block. This butter R8 block exists from the 16-point block means division of 16 points into smaller parts for easy execution. The RTL view and simulation results of butter R8 are shown in Figures 17 and 18, respectively.

Figure 17. Even and odd part of butter R8.

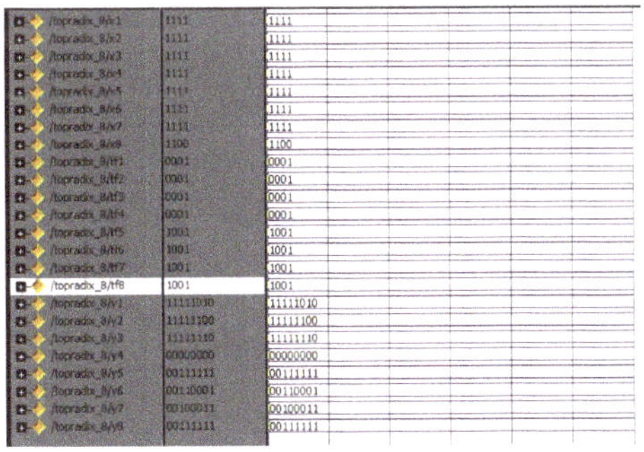

Figure 18. Simulation results of butter R8.

3.1.6. Even and Odd Parts

Even and odd parts are the two different functions in the entire butterfly unit. The combination of even- and odd-part units is present in all modules (M2, M3, M4 and M5). In four modules, four even- and odd-part combinations are present. In the different parts, the divided sequence can be ordered into even and odd parts/places to save memory requirements. The even and odd parts have two instances internally to make the execution easier. The simulation results are shown in Figure 19.

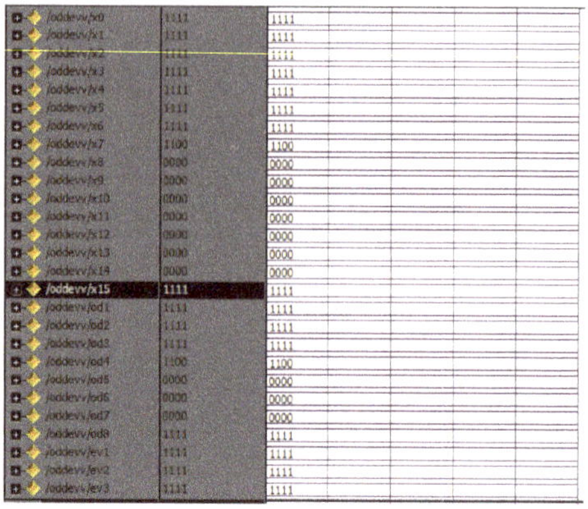

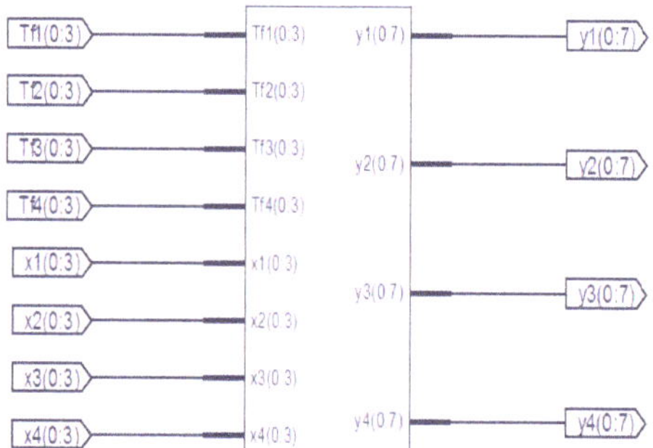

Figure 19. Simulation results of even and odd part of butter R8.

3.1.7. Butter R4

Butter R4 is the basic unit in this structure because it represents the radix-4 function. Each stage has this unit, either as a single unit or as a set/group. In this unit, four inputs and four outputs are present, as shown in the Figure 14. Each input is multiplied with the twiddle factor and gives the output, implying that for the 256-bit input, there are 256 twiddle factors present. The RTL view and simulation results of butter R4 are shown in Figures 20 and 21, respectively.

Figure 20. Butterfly R4 with four twiddle factors.

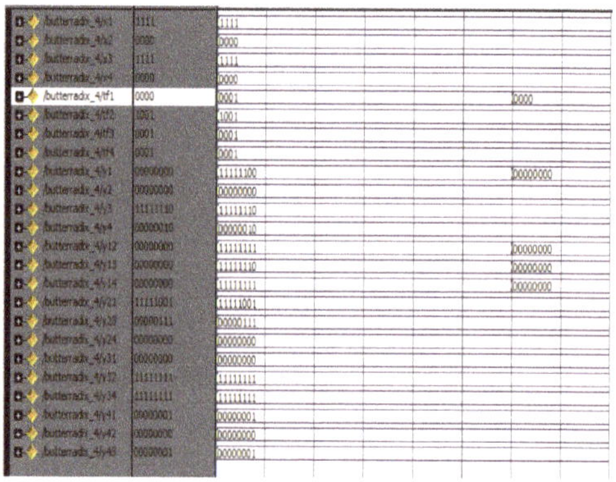

Figure 21. Simulation results of radix-4.

3.1.8. Internal Structure of Butter R4

The internal structure of butter R4 consists of different units. These units are responsible for the entire functionality of the butterfly unit. The different units are adders, subtractors and multipliers, and these units are called basic building blocks for butter R4. The RTL view is shown in Figure 22.

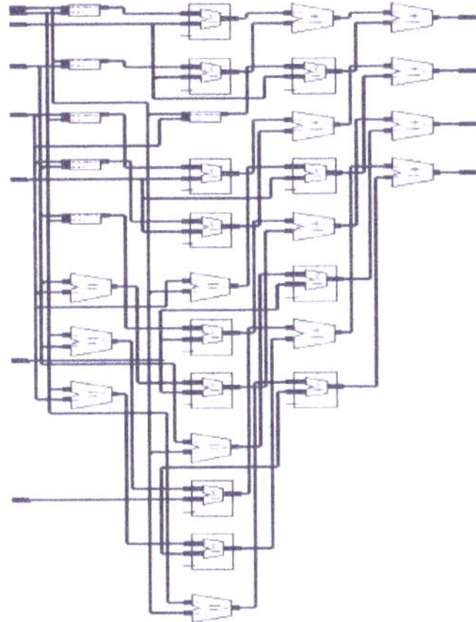

Figure 22. Gate-level structure of R4.

3.2. Simulation Results of Radix-4 Algorithm

Figure 23 shows the simulation results of the 64-point radix-4 DITFFT butterfly block. The A of each of the 4 bits, totally, is 256 bits. W of 256 points represents the inputs and twiddle factors. The X is an output of 64 points for each point of 8 bits, since the even and odd part results, totally, in 512 points, all in binary formats.

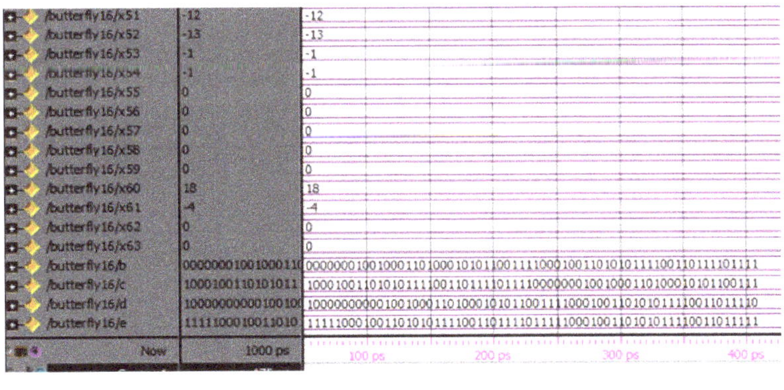

Figure 23. Simulation results of 64-point radix-4 DITFFT.

3.3. Timing Report

The timing report is generated under a speed grade of -5. This report includes all the input and output cells and their fan out. Each gate delay and net delay is also considered, and the summation of gate delay gives the timing delay of the project as 19 ns. The comparison between the performance of the radix-2 and radix-4 algorithms based on different aspects, such as number of slices used, LUTS, bonded IOBs, flip flops and global clocks, for their operations.

From Table 3, it is observed that the minimum delay for the functioning of inputs and outputs for radix-4 is significantly less when compared with radix-2, and memory usage is almost the same as radix-2 and even radix-4 using a greater number of inputs. We observed that 75% of the computations were saved in radix-4, even though the device utilization was higher. Table 4 shows the performance comparison of size, delay and power of previous existing methods.

Table 3. Performance comparison between the radix-2 and radix-4 algorithms.

Device Utilization	Radix-2			Radix-4		
	Used	Available	Utilization (%)	Used	Available	Utilization (%)
Slices	2388	4656	51	3332	4656	71
4 input LUTs	4282	9312	45	5936	9312	63
IOBs	135	232	58	1024	232	441
Delay	75.050 ns			18.963 ns		
Memory	0.206720 GB			0.228 GB		
Power	56.78 mW			12.68 mW		

Table 4. Performance comparison among the proposed FFT with existing works.

Parameter	This Work	[2]	[8]	[10]	[14]
FFT Size	64-4	32-8	64-4	16-4	16-4
Delay (ns)	18.96	419	8.10	2.2	2.67
Power (mW)	12.68	739.5	33.5	3.5	4

3.4. Medical Image Compression

The proposed algorithm for image compression is simulated using the same targeted device, given in Table 3, used for radix-4. The considered medical image was compressed using different tolerance values, such as 0.0007625, 0.003246, 0.013075 and 0.03924, and it also returns the drop values calculated using the formula given in Equation (7) as 0.10, 0.31, 0.61 and 0.83, respectively, as shown in Figures 24 and 25.

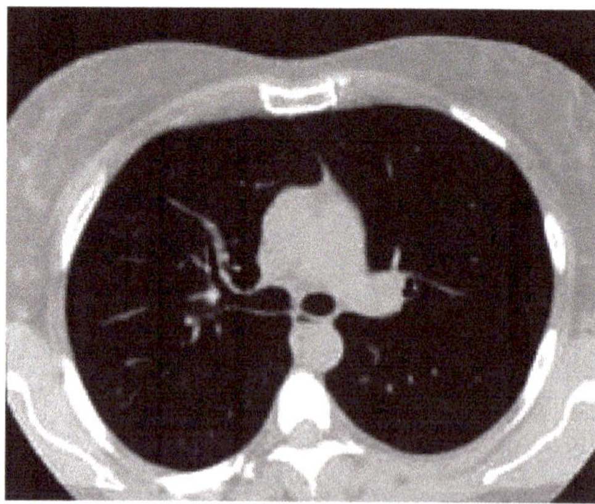

Figure 24. Original image.

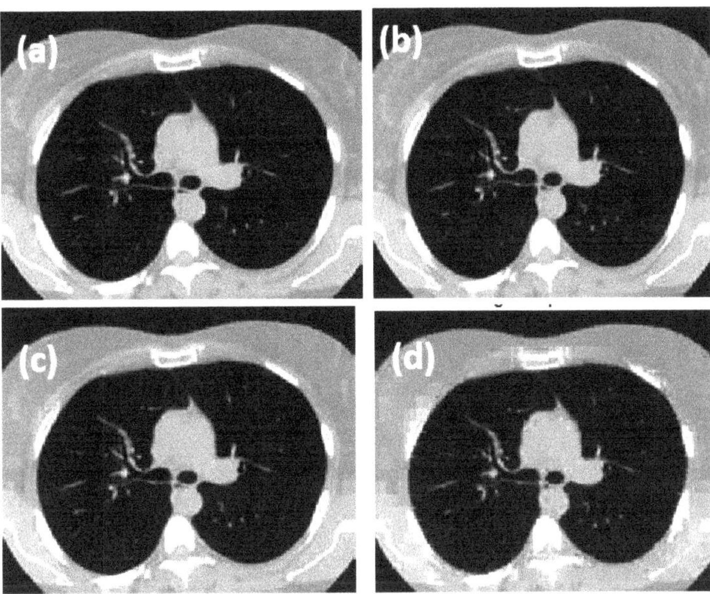

Figure 25. Compressed image with (**a**) tolerance = 0.0007625 resulting in a drop ratio of 0.10.; (**b**) tolerance = 0.003246 resulting in drop ratio of 0.31; (**c**) tolerance = 0.013075 resulting in drop ratio of 0.61; (**d**) tolerance = 0.03924 resulting in a drop ratio of 0.83.

4. Conclusions

In this article, a new high-speed DIT-FFT algorithm based on the radix-4 algorithm for medical image compression was proposed and simulated on a target device xc3s500e-5fg320. The simulation results show that radix-4 processes the input with less delay. From the time delay table, it is very clear that approximately 75% of the processing time is saved with less memory usage. The proposed radix algorithm also showed lower power consumption than the existing radix2, which allows the use of the present algorithm in the medical field, where low-power devices are preferable. This is another milestone for this article. Due to these advantages, the proposed algorithm is used in medical image compression. We observed different tolerances and their drop ratios. It is observed that the scaling factor for image compression depends on tolerance as directly proportional. According to the obtained results for different tolerances, such as 0.0007625, 0.003246, 0.013075 and 0.03924, the drop values are 0.10, 0.31, 0.61 and 0.83, respectively. The level of compression and tolerance values for medical images can be chosen based on the drop ratio and application. Before being taken for ASIC, it can be tested on FPGA for speed and further embedding of the required components.

Author Contributions: All three authors equally contributed in the simulation. All authors have read and agreed to the published version of the manuscript.

Funding: This research received no external funding.

Institutional Review Board Statement: Not Applicable.

Informed Consent Statement: Not Applicable.

Data Availability Statement: Not Applicable.

Conflicts of Interest: The authors declare no conflict of interest.

References

1. Shashikala, B.N.; Sudha, B.S.; Sarkar, S. Efficient Implementation of Radix-2 FFT Architecture using CORDIC for Signal Processing Applications. In Proceedings of the 2020 International Conference on Recent Trends on Electronics, Information, Communication & Technology (RTEICT) IEEE, Bangalore, India, 12–13 November 2020; pp. 137–142.
2. Jyotsna, Y.; Nithiyameenatchi, N.; Konguvel, E.; Kannan, M. Performance analysis of radix-2/3/5 decompositions in fixed point DIT FFT algorithms. In Proceedings of the International Conference on Computer Communication and Informatics (ICCCI), Coimbatore, India, 22–24 January 2020; pp. 1–7.
3. Ganguly, A.; Chakraborty, A.; Banerjee, A. A novel VLSI design of radix-4 DFT in current mode. *Int. J. Electron.* **2019**, *106*, 1845–1863. [CrossRef]
4. Patil, S.D.; Sharma, M. A 2048-point Split-Radix Fast Fourier Transform Computed using Radix-4 Butterfly Units. *Int. J. Recent Technol. Eng. (IJRTE)* **2019**, *8*, 2043–2046. [CrossRef]
5. Ramprabu, G.; Rajmohan, V.; Prakash, V.R.; Shankar, N. Analysis of Feed forward Radix-2^2 FFT 4-Parallel Architecture. In Proceedings of the International Conference on Smart Systems and Inventive Technology (ICSSIT), Tirunelveli, India, 27–29 November 2019; pp. 168–172.
6. Noor, S.M.; John, E.; Panday, M. Design and Implementation of an Ultralow-Energy FFT ASIC for Processing ECG in Cardiac Pacemakers. *IEEE Trans. Very Large-Scale Integr. (VLSI) Syst.* **2019**, *27*, 983–987. [CrossRef] [PubMed]
7. Abbas, Z.A.; Sulaiman, N.B.; Yunus, N.A.M.; Hasan, W.Z.W.; Ahmed, M.K. An FPGA implementation and performance analysis between Radix-2 and Radix-4 of 4096-point FFT. In Proceedings of the IEEE 5th International Conference on Smart Instrumentation, Measurement and Application (ICSIMA), Songkhla, Thailand, 28–30 November 2018; pp. 1–4.
8. Anitha, T.G.; KVijayalakshmi, F.F.T. Based Compression Approach for Medical Images. *Int. J. Appl. Eng. Res.* **2018**, *13*, 3550–3567.
9. Gálvez, M.G.; Sanchez, M.A.; Lopez-Vallejo, M.L.; Grajal, J. A 4096-Point Radix-4 Memory-Based FFT Using DSP Slices. *IEEE Trans. Very Large-Scale Integr. (VLSI) Syst.* **2017**, *25*, 375–379.
10. Mohapatra, B.N.; Mohapatra, R.K. FFT and sparse FFT techniques and applications. In Proceedings of the Fourteenth International Conference on Wireless and Optical Communications Networks (WOCN), Mumbai, India, 24–26 February 2017; pp. 1–5.
11. Neuenfeld, R.H.; Fonseca, M.B.; da Costa, E.A.C.; Oses, J.P. Exploiting addition schemes for the improvement of optimized radix-2 and radix-4 FFT butterflies. In Proceedings of the IEEE 8th Latin American Symposium on Circuits & Systems (LASCAS), Bariloche, Argentina, 20–23 February 2017; pp. 1–4.
12. Anitha, T.G.; Vijayalakshmi, K. Design of Novel FFT Based Image Compression Algorithms and Architectures. *Int. J. Progress. Sci. Technol. (IJPSAT)* **2017**, *5*, 24–42.
13. Walia, S.; Majithia, S. Adaptive Gaussian Filter Based Image Recovery Using Local Segmentation. *Int. J. Technol. Comput. (IJTC)* **2016**, *2*, 163–172.
14. Neuenfeld, R.; Fonseca, M.; Costa, E. Design of optimized radix-2 and radix-4 butterflies from FFT with decimation in time. In Proceedings of the IEEE 7th Latin American Symposium on Circuits & Systems (LASCAS), Florianopolis, Brazil, 28 February–2 March 2016; pp. 171–174.
15. Qian, Z.; Margala, M. Low-Power Split-Radix FFT Processors Using Radix-2 Butterfly Units. *IEEE Trans. Very Large-Scale Integr. (VLSI) Syst.* **2016**, *24*, 3008–3012. [CrossRef]
16. Ma, Z.-G.; Yin, X.-B.; Yu, F. A novel memory-based FFT architecture for real-valued signals based on a radix-2 decimation-in-frequency algorithm. *IEEE Trans. Circuits Syst. II Express Briefs* **2015**, *62*, 876–880. [CrossRef]
17. Jayaram, K.; Arun, C. Survey report for Radix-2, Radix-4 and Radix-8 FFT Algorithms. *Int. J. Innov. Res. Sci. Eng. Technol.* **2015**, *4*, 5149–5154.
18. Brundavani, P. FPGA Implementation of 256-Bit, 64-Point DIT-FFT Using Radix-4 Algorithm. *Int. J. Adv. Res. Comput. Sci. Softw. Eng.* **2015**, *3*, 126–133.
19. Qian, Z.; Nasiri, N.; Segal, O.; Margala, M. FPGA implementation of low-power split-radix fast fourier transform processors. In Proceedings of the 24th International Conference, Field Program, Logic Applications, Munich, Germany, 2–4 September 2014; pp. 1–2.
20. Reddy, A.; Suman, V. Design and Simulation of FFT Processor Using Radix-4 Algorithm Using FPGA. *Int. J. Adv. Sci. Technol.* **2013**, *61*, 53–62. [CrossRef]

Proceeding Paper

Lensless Hyperspectral Phase Retrieval via Alternating Direction Method of Multipliers and Spectral Proximity Operators [†]

Igor Shevkunov *, Vladimir Katkovnik and Karen Egiazarian

Computational Imaging Group, Faculty of Information Technology and Communication Sciences, Tampere University, 33720 Tampere, Finland; vladimir.katkovnik@tuni.fi (V.K.); karen.eguiazarian@tuni.fi (K.E.)

* Correspondence: igor.shevkunov@tuni.fi

† Presented at the International Conference on "Holography Meets Advanced Manufacturing", Online, 20–22 February 2023.

Abstract: We consider the application of a recently developed hyperspectral broadband phase retrieval (HSPhR) technique for spectrally varying object and modulation phase masks at 100 spectral components. The HSPhR method utilizes advanced techniques such as Spectral Proximity Operators and ADMM to retrieve complex-domain spectral components from multiple spectral observations. These techniques filter out noisy observations and strike a balance between noisy intensity observations and their predicted counterparts, resulting in accurate retrieval of the broadband hyperspectral phase even for low signal-to-noise ratio components. Both simulation and physical experiments have confirmed the effectiveness of this approach.

Keywords: hyperspectral imaging; sparse representation; noise filtering; phase imaging

1. Introduction

Hyperspectral imaging (HSI) is a well-known technique for spectral observations. It is used in various applications, e.g., earth surface remote sensing [1], and medical and bio-medical diagnostics [2]. HSI retrieves information based on images obtained across a wide spectrum range with hundreds to thousands of spectral channels. These images are stacked in 3D cubes, where the first two dimensions are spatial coordinates, and the third dimension is a spectral channel, represented by either wavenumber or wavelength.

Phase HSI is a special class of HSI where images of interest are complex-valued, and both phase and amplitude are visualized. It is a promising technique because the amount of retrieved information is doubled compared to real-valued HSI. It goes directly from the nature of complex-domain imaging, where each image is complex-valued with phase and amplitude. Phase brings information about the light delay refracted/transmitted from/through an object. This delay might be recalculated into valuable information, e.g., dry mass [3], refractive index [4], or thickness [5].

Recently, HS digital holography has been developed, which, additionally to conventional holography, is able to recover spectrally resolved phase/amplitude information (e.g., Refs. [6,7]). In our previous papers [8–10], the phase HSI formulation was done for the observation model provided that the parametric model of the object is known and has been used for phase delay recalculation between spectral channels. As a result, the HS absolute phases of the object are reconstructed. In the papers [8–10], we used an observation model where separate diffraction patterns were registered for separate wavelengths and masks. More close papers are [11,12], where we based our algorithms on the Fourier spectroscopy. In Ref. [11], we used the model for the object with the connection between the spectral phases through the thickness. However, the proposed new model for phase HSI is different from the Fourier spectroscopy approach as it does not use any instruments for spectral analysis, such as the harmonic reference beams and precise scanning phase delay stages [6,7].

Citation: Shevkunov, I.; Katkovnik, V.; Egiazarian, K. Lensless Hyperspectral Phase Retrieval via Alternating Direction Method of Multipliers and Spectral Proximity Operators. *Eng. Proc.* **2023**, *34*, 19. https://doi.org/10.3390/HMAM2-14146

Academic Editor: Vijayakumar Anand

Published: 13 March 2023

The spectral resolution for the considered approach is based on the diversity of spectral properties of the image formation operators and masks [13].

In this paper, we consider applying the developed in Ref. [13] HS phase retrieval algorithm for an object whose phase and amplitude are characterized on a spectrum with 100 spectral components in simulation and physical experiments.

2. Hyperspectral Phase Imaging

In this section, we briefly describe the HS phase retrieval algorithm developed in Ref. [13]; for a detailed derivation of the solutions, please follow the pioneering paper [13].

To solve the lensless phase retrieval problem, we utilize random phase masks $\mathcal{M}_{t,k} \in \mathbb{C}^{T \times N}$ which, along with propagation operator $A_{t,k}$, encode information about an object $U_{o,k}$ into the observations Y_t [14]:

$$Y_t = \sum_{k \in K} |A_{t,k}(\mathcal{M}_{t,k} \circ U_{o,k})|^2, \ t = 1, \ldots, T, \tag{1}$$

where $Y_t \in \mathbb{R}^N$, and $A_{t,k} \in \mathbb{C}^{K \times N}$ is an image formation operator modeling propagation of 2D object images from the object plane to the sensor, and '$\circ$' stands for the element-by-element product of two vectors. For the object of interest, it is the vector $U_{o,k} \in \mathbb{C}^{K \times N}$, $N = nm$, where n and m are the width and height of 2D image; k stays for the spectral variable. The HS phase retrieval is a reconstruction of a complex-valued object $U_{o,k} \in \mathbb{C}^{K \times N}$, $k \in K$, from these intensity measurements Y_t (1). For essential noisy observations with additive noise ε_t, observations Y_t are complemented by noise ε_t:

$$Z_t = Y_t + \varepsilon_t, \ t \in T. \tag{2}$$

The total intensity measurement $Y_t \in R^N$ is a sum of intensities from all spectral components K, which means that spectral information is severely mixed in the observations.

2.1. HS Phase Retrieval Solution (HSPhR)

Traditionally, the solution $U_{o,k}$ is found by measuring the mismatch between the observations and the prediction of the intensities of $U_{t,k}$ summarized over the spectral interval. We realize it through a neg-log-likelihood of the observed $\{Z_t\}$, $t \in T$, as $l(\{Z_t\}, \sum_{k \in K} |U_{t,k}|^2)$, where $U_{t,k}$ are the complex-valued object wavefronts at the sensor plane calculated as $U_{t,k} = A_{t,k}(\mathcal{M}_{t,k} \circ U_{o,k})$. In the following text, the curly brackets signify a group of variables. We use an unconstrained maximum likelihood optimization to reconstruct the object 3D cube $\{U_{o,k}\}$, $k \in K$, from the criterion of the form:

$$\min_{\{U_{o,k}\}} l(\{Z_t\}, \sum_{k \in K} |A_{t,k}(\mathcal{M}_{t,k} \circ U_{o,k})|^2) + f_{reg}(\{U_{o,k}\}), \tag{3}$$

where a second summand is an object prior. We solve it by alternating direction method of multipliers (ADMM) [15–18]. Next, Equation (3) is reformulated as a constrained optimization:

$$\min_{\{U_{t,k}, U_{o,k}\}} l(\{Z_t\}, \sum_{k \in K} |U_{t,k}|^2) + f_{reg}(\{U_{o,k}\}), \tag{4}$$

$$\text{w.s.t. } U_{t,k} = A_{t,k}(\mathcal{M}_{t,k} \circ U_{o,k}).$$

We resolve (4) by the unconstrained formulation with the parameter $\gamma > 0$:

$$J = l(Z_t, \sum_k |U_{t,k}|^2) + \frac{1}{\gamma} \sum_{t,k} ||U_{t,k} - A_{t,k}(\mathcal{M}_{t,k} \circ U_{o,k})||_2^2 + f_{reg}(\{U_{o,k}\}). \tag{5}$$

The second summand is the quadratic penalty for the difference between $A_{t,k}(\mathcal{M}_{t,k} \circ U_{o,k})$ and the splitting $U_{t,k}$. In the optimization of (5), $U_{t,k} \rightarrow A_{t,k}(\mathcal{M}_{t,k} \circ U_{o,k})$ as $\gamma \rightarrow 0$. The minimization algorithm iterates $\min_{\{U_{t,k}\}} J$ with given $U_{o,k}$ and $\min_{\{U_{o,k}\}} J$ with given $U_{t,k}$:

$$U_{t,k}^{(s+1)} = \arg\min_{U_{t,k}} (l(Z_t, \sum_k |U_{t,k}|^2) + \frac{1}{\gamma} \sum_{t,k} ||U_{t,k} - A_{t,k}(\mathcal{M}_{t,k} \circ U_{o,k})^{(s)}||_2^2), \tag{6}$$

$$U_{o,k}^{(s+1)} = \arg\min_{U_{o,k}} (f_{reg}(\{U_{o,k}\}) + \frac{1}{\gamma} \sum_{t,k} ||U_{t,k}^{(s+1)} - A_{t,k}(\mathcal{M}_{t,k} \circ U_{o,k})||_2^2). \tag{7}$$

For the iterative solution of (6) and (7), we introduce the Lagrangian variables $\Lambda_{t,k}$ [19]:

$$U_{t,k}^{(s+1)} = \arg\min_{U_{t,k}} J(\{U_{t,k}, U_{o,k}^{(s)}, \Lambda_{t,k}^{(s)}\}),$$

$$U_{o,k}^{(s+1)} = \arg\min_{U_{o,k}} J(\{U_{t,k}^{(s+1)}, U_{o,k}, \Lambda_{t,k}^{(s)}\}), \tag{8}$$

$$\Lambda_{t,k}^{(s+1)} = \Lambda_{t,k}^{(s)} - \frac{1}{\gamma}(U_{t,k}^{(s+1)} - A_{t,k}(\mathcal{M}_{t,k} \circ U_{o,k}^{(s+1)}),$$

where s is the iterative variable.

Minimizing (7) on $U_{o,k}$, we replace the regularization term $f_{reg}(\{U_{o,k}, k \in K\})$ with noise suppression in $\{U_{o,k}, k \in K\}$. It is done by Complex Cube Filter (CCF) [20], developed specifically for 3D hyperspectral complex-domain images. Then, the solution for $U_{o,k}$ is of the form

$$U_{o,k} = \frac{\sum_t A_{t,k}^H \mathcal{M}_{t,k}^H (U_{t,k} - \Lambda_{t,k})}{\sum_t A_{t,k}^H A_{t,k} + reg}, \tag{9}$$

where the regularization parameter $reg > 0$ is included if $\sum_t A_{t,k}^H A_{t,k}$ is singular or ill-conditioned.

Minimization on $U_{t,k}$ (6) depends on the noise type in observations $\{Z_t\}$. We consider two types of noise: Poisson and Gaussian. For the Gaussian noise, the loss function is

$$J = \frac{1}{\sigma^2} \sum_t ||Z_t - \sum_k |U_{t,k}|^2||_2^2 + \frac{1}{\gamma} \sum_{t,k} ||U_{t,k} - A_{t,k}(\mathcal{M}_{t,k} \circ U_{o,k}) - \Lambda_{t,k}||_2^2. \tag{10}$$

And for the Poisson noise, considering its multiplicative nature, the loss function is the following.

$$J = \sum_t (\chi \sum_k |U_{t,k}|^2 - Z_t \log(\sum_k |U_{t,k}|^2 \chi)) + \tag{11}$$

$$+ \frac{1}{\gamma} \sum_{t,k} ||U_{t,k} - A_{t,k}(\mathcal{M}_{t,k} \circ U_{o,k}) - \Lambda_{t,k}||_2^2 + f_{reg}(\{U_{o,k}\}_{k \in K}),$$

where $\chi > 0$ is a scaling parameter for photon flow, it is proportional to the camera exposure time and defines the noise level of a signal. The signal-to-noise ratio $E\{Z_t\}^2 / var\{Z_t\} = Y_t \chi$ takes larger values for larger χ.

To minimize $\min_{\{U_{t,k}\}} J$, we need to calculate the derivatives with respect to $U_{t,k}^*$ and then evaluate the necessary minimum conditions by setting the derivative to zero. This process involves solving a set of complex-valued cubic equations for Gaussian observations (10):

$$[\frac{2}{\sigma^2}(\sum_{k \in K} |U_{t,k}|^2 - Z_t) + \frac{1}{\gamma}] U_{t,k} = \frac{1}{\gamma}(A_{t,k}(\mathcal{M}_{t,k} \circ U_{o,k}) + \Lambda_{t,k}). \tag{12}$$

With a solution in the form

$$\hat{U}_{t,k} = \frac{A_{t,k}(\mathcal{M}_{t,k} \circ U_{o,k}) + \Lambda_{t,k}}{1 + \frac{2\gamma}{\sigma^2}\left(\sum_{k \in K} |\hat{U}_{t,k}|^2 - Z_t\right)}.$$ (13)

For Poisson observations (11), minimization leads to the set of non-linear equations with respect to $U_{t,k}$:

$$\left[\chi - \frac{Z_t}{\sum_{k \in K} |U_{t,k}|^2} + \frac{1}{\gamma}\right] U_{t,k} = \frac{1}{\gamma}\left(A_{t,k}(\mathcal{M}_{t,k} \circ U_{o,k}) + \Lambda_{t,k}\right).$$ (14)

With a solution in the form

$$\hat{U}_{t,k} = \frac{A_{t,k}(\mathcal{M}_{t,k} \circ U_{o,k}) + \Lambda_{t,k}}{1 + \gamma\chi + \frac{\gamma Z_t}{\sum_{k \in K} |U_{t,k}|^2}}, \, k \in K.$$ (15)

The solutions for $\min_{\{U_{t,k}\}} J$ can be interpreted as spectral proximity operators (SPO), generated by minimizing the likelihood items with a quadratic penalty, for both Gaussian and Poisson:

$$\hat{U}_{t,k} = prox_{f\gamma}(A_{t,k}(\mathcal{M}_{t,k} \circ U_{o,k}) + \Lambda_{t,k}),$$ (16)

where f stays for the minus-log-likelihood part of J and $\gamma > 0$ is a parameter.

These operators solve two problems: First, they extract and separate complex-valued spectral components $U_{t,k}$ from the real-valued observations Z_t, in which these components are mixed into the total intensity of the signal. Thus, it provides the spectral analysis of the signals. Second, the noisy observations Z_t are filtered with the power controlled by the parameter γ compromising Z_t and the power of the predicted signal $A_{t,k}(\mathcal{M}_{t,k} \circ U_{o,k})$ at the sensor plane. For a detailed derivation of the solutions, follow the pioneering paper [13].

2.2. Developed Algorithm

A block scheme of the algorithm is shown in Figure 1. Complex domain initialization (Step 1) is required for the considered spectral domain $k \in K$. In our experiments, we make a 2D random white-noise Gaussian distribution for phase and a uniform 2D positive distribution on (0,1] for amplitude, which are independent for each k. The Lagrange multipliers are initialized by zero values, $\Lambda_{t,k} = 0$. The forward propagation is realized through the angular spectrum approach and produced for all $k \in K$ and $t \in T$ (Step 2). The update of the wavefront at the sensor plane (Step 3) is produced by the proximal operators (SPO). For the Gaussian observations, this operator is defined by (13), and for the Poisson observations—by (15). Calculating $\sum_{k \in K} |\hat{U}_{t,k}|^2$ requires solving the polynomial equations, cubic or quadratic, for the Gaussian and Poisson cases, respectively. In Step 4, the Lagrange variables are updated. The backward propagation of the wavefront from the sensor plane to the object plane is combined with an update of the spectral object estimate in Step 5. The sparsity-based regularization by Complex Cube Filter (CCF) [20] is relaxed by the weight-parameter $0 < \beta < 1$ at Step 6. The iteration number is fixed to n.

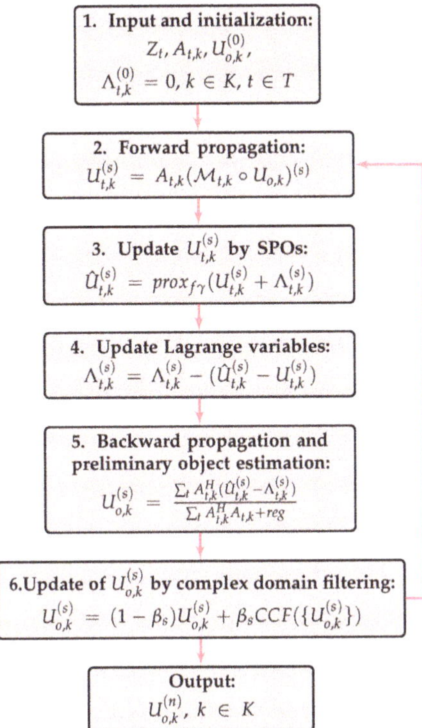

Figure 1. HSPhR algorithm. The total number of iterations is n.

3. Numerical Experiments

Simulation experiments are produced for the complex-valued wavefronts obtained from the propagation of an HS light beam through a thin transparent object. We define the phase delay of the object as:

$$\varphi_k = \frac{\phi \cdot \lambda_1}{\lambda_k} \frac{n_{\lambda_k} - 1}{n_{\lambda_1} - 1}, k \in [1:K] \tag{17}$$

where λ_k is a wavelength of the wavefront ($\Lambda = 550$–950 nm) with total number of wavelengths $K = 100$, ϕ is a basic phase distribution, and n_{λ_k} is the refractive index of the optical material of the object. We took the phase distribution for ϕ as a vortex beam phase in the range of $[-\pi, \pi]$ and amplitude as a logo of Tampere University in the range of $[0, 1]$, see Figure 2. Spectral amplitude dependence is modeled in accordance with the spectrum of a supercontinuum white laser (see Section 4), providing a non-uniform spectral distribution. This non-uniformity strongly complicates the reconstruction process because of low-intensity spectral components.

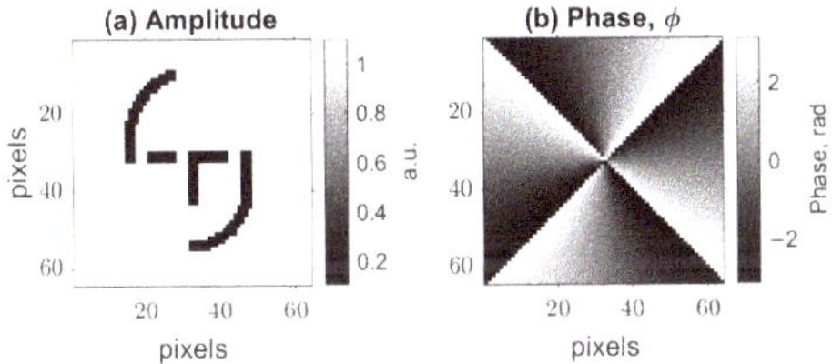

Figure 2. Modeled amplitude and basic phase ϕ of the HS object.

It is of particular interest to investigate the dependence between the accuracy of the algorithm and the wavelength number, represented by K, as well as the number of observations denoted by T. To evaluate the accuracy of the complex-valued reconstruction, we use the relative error criterion [21]:

$$ERROR_{rel} = \min_{\varphi \in [0,2\pi)} \frac{||\hat{x} \circ \exp(j\varphi) - x||_2^2}{||x||_2^2}, \qquad (18)$$

where x and $\hat{x}$ are the true signal and its estimate. If $ERROR_{rel}$ is less than 0.1 the quality of imaging is high. We calculate $ERROR_{rel}$ for $K = [6, 20, 60, 100]$ and $T = [18, 60, 180, 300]$. The wavelengths for the varying K are distributed uniformly covering the spectral interval $[550, 950]$ nm. The number of iterations is fixed to $n = 200$. The relative errors $ERROR_{rel}$ obtained in these experiments are shown in Figure 3a. It is shown that $ERROR_{rel}$ is low for the high number of masks, T, and for high-quality imaging, the needed number of masks equals the doubled number of wavelengths. Therefore, for the simulation case of $K = 100$ we took $T = 200$.

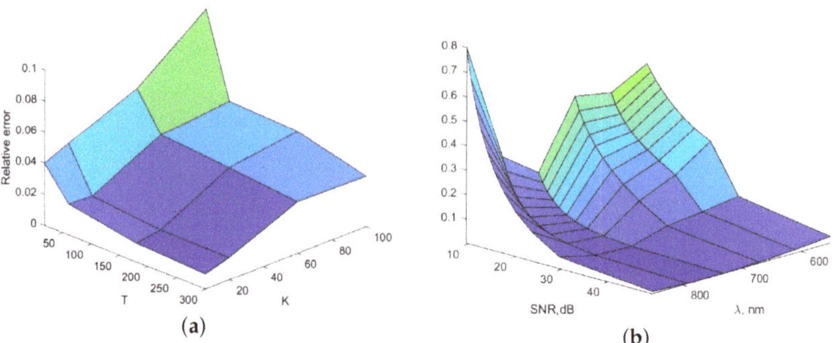

(a)   (b)

Figure 3. HSPhR reconstruction relative error maps. (**a**) Mean relative errors depending on the number of masks, T, and the number of wavelengths, K. A number of iterations $n = 200$, SNR = 54 dB. (**b**) Dependence of relative errors on the SNR and wavelength.

In Figure 3b, we show $ERROR_{rel}$ dependence on the noise level in the observations, which is characterized by the signal-to-noise ratio (SNR) in dB and on wavelength with $K = 6$ and $T = 18$. It is seen that for noise levels with SNR bigger than 30 dB $ERROR_{rel}$ is low and indicates high-quality imaging. It is interesting to note that for high noise levels (SNR < 30 dB) the shape of $ERROR_{rel}$ corresponds to the inverted spectral distribution of the modeled illumination: the spectral components with higher values (middle of the spectrum) provide lower $ERROR_{rel}$ than the spectral components with low values.

In Figure 4, we show a contour $ERROR_{rel}$ map for the HSPhR algorithm for all spectral components (Y-axis) during iterative calculations (X-axis for iteration number). High-quality imaging regions ($ERROR_{rel} < 0.1$) are dark blue on the map with signed contours '0.1'. At first iterations, high-quality imaging appears at only high-intensity spectral components, but with the growing number of iterations, high-quality spreads to low-intensity spectral components. Similar behavior for the spectrally dependent object was demonstrated in Ref. [20], where the developed filter CCF retrieved data from extremely noisy spectral components. Therefore, it is an illustration of the computational overcoming of Fellgett's disadvantage [22] typical for HS observations.

The corresponding reconstruction results after the last iteration ($n = 200$) are presented in Figure 5, where object phase and amplitude are spectrally resolved and correspond to the modeled object from Figure 2.

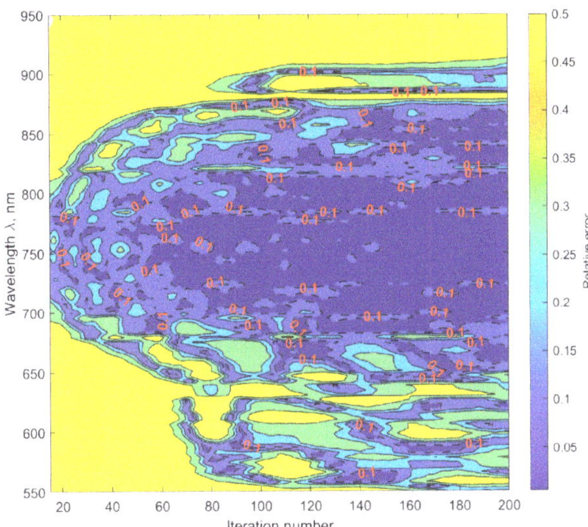

Figure 4. Contour $ERROR_{rel}$ map for HSPhR reconstruction. $K = 100$, $T = 200$, SNR = 54 dB. Dark blue regions correspond to high-quality imaging.

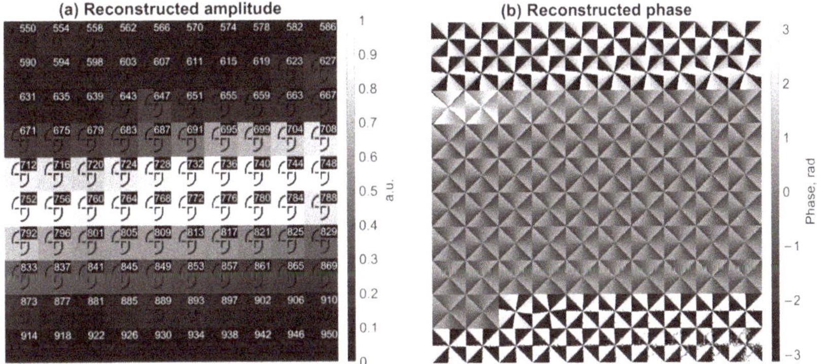

Figure 5. Reconstructed amplitudes (**a**) and phases (**b**) of the object in simulations. Reconstruction is performed for $K = 100$ wavelengths with $T = 200$ masks and after $n = 200$ iterations. The wavelength value is written in each image of the amplitude and corresponds to phase images with the same location.

4. Experimental Validation

For the experimental validation, we used a setup with a phase-only spatial light modulator (SLM) and supercontinuum laser source, see Figure 6. A sum of the object $U_{o,k}$ and masks $\mathcal{M}_{t,k}$ was imaged on SLM, and projected to the 'Object' plane by achromatic lenses 'L_3' and 'L_4'. The light wavefront then propagates freely on distance $d = 2.2$ mm from the 'Object' plane to 'CMOS', where it is registered as a noisy intensity distribution, Z_t, with $t \in [1, T]$. According to the simulation results, the mask spatial distribution was modeled as piecewise invariant random with equal probabilities taking one of the following five values $[0, 1, 0, 0.25, 0.75] \cdot pi$, for the $\lambda = 520$ nm. The object was taken as an image of a cameraman with 64×64 pixels, and phase images of a single randomly picked mask and object are in Figure 7. SLM is the Holoeye phase-only GAEA-2-vis panel, resolution 4160×2464, pixel size 3.74 μm. The laser is YSL photonics CS-5 with $\Lambda = 470 \div 2400$ nm. To work in the spectral range of the sensor, we limit the laser's spectral bandwidth to the

$\Lambda = 470 \div 1000$ nm range by a bandpass filter. The camera is a monochrome Blackfly S board with the matrix Sony IMX264, 3.45 µm pixels, and 2448×2048 pixels.

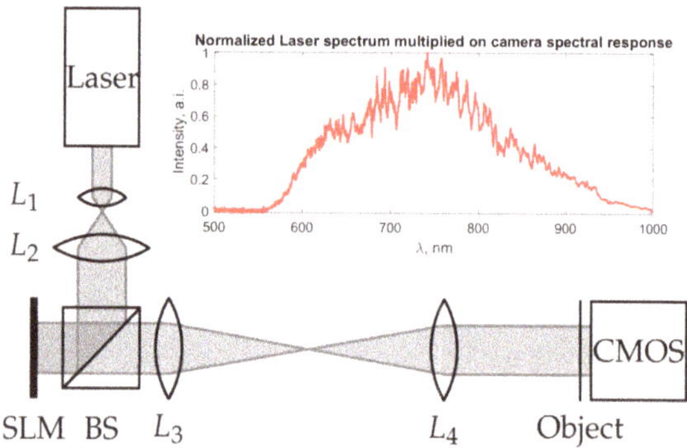

Figure 6. Optical setup. The laser is a supercontinuum light source with wavelengths in the range of 550–1000 nm, L_1, L_2 are beam-expanding lenses, 'BS' is a beamsplitter, 'SLM' is a Spatial Light Modulator, L_3, L_4 are lenses in a 4f-telescopic system, 'Object' is the plane of the projected phase object and masks from SLM, and CMOS is a camera.

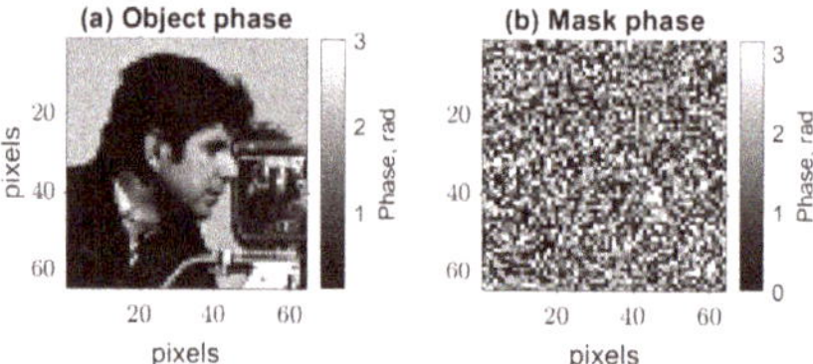

Figure 7. Images of object (**a**) and mask (**b**) imaged on SLM in experiments for $\lambda = 520$ nm.

The reconstruction results are presented in Figure 8, which we made from $T = 300$ observations and for $K = 100$ wavelengths. The number of observations is taken $T = 300$ to overcome noise problems since the estimated [23] SNR of observations equals 34 dB, which is close to the lower limit of HSPhR, estimated in the simulations. Reconstructed amplitude intensities correspond to the spectral distribution of the used laser with a maximum at $\lambda = 750$ nm, and phase images correspond to the given cameraman image. Image quality varies from low to high with respect to the intensity distribution of spectral components.

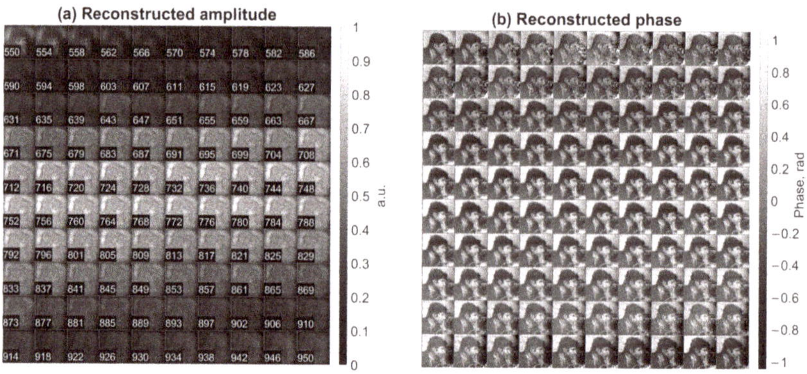

Figure 8. Experimentally reconstructed amplitudes (**a**) and phases (**b**) of the object. Reconstruction is performed for $K = 100$ wavelengths with $T = 300$ masks. The wavelength number is written in each image of the amplitude and corresponds to phase images with the same location.

5. Conclusions

We have considered an application of an approach for hyperspectral phase retrieval that utilizes modulation phase masks for the reconstruction of a spectrally variable object with 100 spectrally dependent components. The approach which is based on a complex domain version of the ADMM method and spectral proximity operators successfully solved phase retrieval problem in the considered scenario. The proposed technique is able to retrieve complex-domain spectral components of an object from noisy observations and filter out noise by compromising between noisy intensity observations and their predicted counterparts. In the algorithm, we do not use traditional phase retrieval constraints, e.g., known aperture or phase connection among spectral components through the object thickness. With the mask imaged on a transmissive spatial light modulator, the proposed optical implementation allows a simple lensless configuration, which is significantly simpler than traditional hyperspectral approaches, as interferometry and holographic imaging. The proposed approach could potentially be useful in various applications, such as biomedical imaging, remote sensing, and materials science.

Author Contributions: I.S. developed the algorithm, conceived and designed the experiments; V.K. developed the approach and algorithm; K.E. supervised the project. All authors have read and agreed to the published version of the manuscript.

Funding: I.S. was supported by the Academy of Finland (project no. 336357, PROFI 6—TAU Imaging Research Platform).

Institutional Review Board Statement: Not applicable.

Informed Consent Statement: Not applicable.

Data Availability Statement: Data are available on request.

Conflicts of Interest: The authors declare no conflict of interest.

Abbreviations

The following abbreviations are used in this manuscript:

BS	Beam Splitter
CMOS	Complementary Metal–Oxide–Semiconductor
CCF	Complex Cube Filter
HSI	HyperSpectral Imaging
HSPhR	HyperSpectral Phase Retrieval
HSDH	HyperSpectral Digital Holography
SPO	Spectral Proximity Operator
SNR	Signal-to-Noise Ratio

References

1. Gao, B.C.; Davis, C.; Goetz, A. A review of atmospheric correction techniques for hyperspectral remote sensing of land surfaces and ocean color. In Proceedings of the 2006 IEEE International Symposium on Geoscience and Remote Sensing, Denver, CO, USA, 31 July–4 August 2006; IEEE: New York, NY, USA, 2006; pp. 1979–1981.
2. Calin, M.A.; Parasca, S.V.; Savastru, D.; Manea, D. Hyperspectral imaging in the medical field: Present and future. *Appl. Spectrosc. Rev.* **2014**, *49*, 435–447. [CrossRef]
3. Ghenim, L.; Allier, C.; Obeid, P.; Hervé, L.; Fortin, J.Y.; Balakirev, M.; Gidrol, X. A new ultradian rhythm in mammalian cell dry mass observed by holography. *Sci. Rep.* **2021**, *11*, 1290. [CrossRef] [PubMed]
4. Belashov, A.V.; Zhikhoreva, A.A.; Bespalov, V.G.; Novik, V.I.; Zhilinskaya, N.T.; Semenova, I.V.; Vasyutinskii, O.S. Refractive index distributions in dehydrated cells of human oral cavity epithelium. *J. Opt. Soc. Am. B* **2017**, *34*, 2538–2543. [CrossRef]
5. Cazac, V.; Meshalkin, A.; Achimova, E.; Abashkin, V.; Katkovnik, V.; Shevkunov, I.; Claus, D.; Pedrini, G. Surface relief and refractive index gratings patterned in chalcogenide glasses and studied by off-axis digital holography. *Appl. Opt.* **2018**, *57*, 507. [CrossRef] [PubMed]
6. Kalenkov, S.G.; Kalenkov, G.S.; Shtanko, A.E. Hyperspectral holography: An alternative application of the Fourier transform spectrometer. *J. Opt. Soc. Am. B* **2017**, *34*, B49. [CrossRef]

7. Claus, D.; Pedrini, G.; Buchta, D.; Osten, W. Accuracy enhanced and synthetic wavelength adjustable optical metrology via spectrally resolved digital holography. *J. Opt. Soc. Am. A Opt. Image Sci. Vis.* **2018**, *35*, 546–552. [CrossRef] [PubMed]
8. Katkovnik, V.; Egiazarian, K. Multi-Frequency Phase Retrieval from Noisy Data. In Proceedings of the 2018 26th European Signal Processing Conference (EUSIPCO), Rome, Italy, 3–7 September 2018; IEEE: New York, NY, USA, 2018; pp. 2200–2204. [CrossRef]
9. Katkovnik, V.; Shevkunov, I.; Petrov, N.V.; Eguiazarian, K. Multiwavelength absolute phase retrieval from noisy diffractive patterns: Wavelength multiplexing algorithm. *Appl. Sci.* **2018**, *8*, 719. [CrossRef]
10. Katkovnik, V.; Shevkunov, I.; Petrov, N.V.; Egiazarian, K. Multiwavelength surface contouring from phase-coded noisy diffraction patterns: Wavelength-division optical setup. *Opt. Eng.* **2018**, *57*, 1. [CrossRef]
11. Shevkunov, I.; Katkovnik, V.; Egiazarian, K. Lensless hyperspectral phase imaging in a self-reference setup based on Fourier transform spectroscopy and noise suppression. *Opt. Express* **2020**, *28*, 17944. [CrossRef] [PubMed]
12. Katkovnik, V.; Shevkunov, I.; Egiazarian, K. Broadband Hyperspectral Phase Retrieval From Noisy Data. In Proceedings of the 2020 IEEE International Conference on Image Processing (ICIP), Abu Dhabi, United Arab Emirates, 25–28 October 2020; IEEE: New York, NY, USA, 2020; pp. 3154–3158. [CrossRef]
13. Katkovnik, V.; Shevkunov, I.; Egiazarian, K. ADMM and spectral proximity operators in hyperspectral broadband phase retrieval for quantitative phase imaging. *Signal Process.* **2023**, *210*, 109095. [CrossRef]
14. Katkovnik, V.; Shevkunov, I.; Petrov, N.N.V.; Egiazarian, K. Computational wavelength resolution for in-line lensless holography: Phase-coded diffraction patterns and wavefront group-sparsity. *Proc. SPIE* **2017**, *10335*, 1033509. [CrossRef]
15. Hestenes, M.R. Multiplier and gradient methods. *J. Optim. Theory Appl.* **1969**, *4*, 303–320. [CrossRef]
16. Eckstein, J.; Bertsekas, D.P. On the Douglas—Rachford splitting method and the proximal point algorithm for maximal monotone operators. *Math. Program.* **1992**, *55*, 293–318. [CrossRef]
17. Afonso, M.V.; Bioucas-Dias, J.M.; Figueiredo, M.A.T. Fast Image Recovery Using Variable Splitting and Constrained Optimization. *IEEE Trans. Image Process.* **2010**, *19*, 2345–2356. [CrossRef] [PubMed]
18. Afonso, M.V.; Bioucas-Dias, J.M.; Figueiredo, M.A.T. An Augmented Lagrangian Approach to the Constrained Optimization Formulation of Imaging Inverse Problems. *IEEE Trans. Image Process.* **2011**, *20*, 681–695. [CrossRef] [PubMed]
19. Li, L.; Wang, X.; Wang, G. Alternating Direction Method of Multipliers for Separable Convex Optimization of Real Functions in Complex Variables. *Math. Probl. Eng.* **2015**, *2015*, 104531. [CrossRef]
20. Shevkunov, I.; Katkovnik, V.; Claus, D.; Pedrini, G.; Petrov, N.V.; Egiazarian, K. Hyperspectral phase imaging based on denoising in complex-valued eigensubspace. *Opt. Lasers Eng.* **2020**, *127*, 1–9. [CrossRef]
21. Candes, E.; Wakin, M. An Introduction To Compressive Sampling. *IEEE Signal Process. Mag.* **2008**, *25*, 21–30. [CrossRef]
22. Fellgett, P.B. On the Ultimate Sensitivity and Practical Performance of Radiation Detectors. *J. Opt. Soc. Am.* **1949**, *39*, 970–976. [CrossRef] [PubMed]
23. Mäkitalo, M.; Foi, A. Noise Parameter Mismatch in Variance Stabilization, With an Application to Poisson–Gaussian Noise Estimation. *IEEE Trans. Image Process.* **2014**, *23*, 5348–5359. [CrossRef]

engineering proceedings

Proceeding Paper

Resilient Calcination Transformed Micro-Optics [†]

Darius Gailevicius *, Rokas Zvirblis and Mangirdas Malinauskas *

Laser Research Center, Physics Faculty, Vilnius University, Sauletekio Ave. 10, 10223 Vilnius, Lithuania

* Correspondence: darius.gailevicius@ff.vu.lt (D.G.); mangirdas.malinauskas@ff.vu.lt (M.M.)

[†] Presented at the International Conference on "Holography Meets Advanced Manufacturing", Online, 20–22 February 2023.

Abstract: Three-dimensional multiphoton laser lithography of hybrid resins has been shown to be a viable tool for producing micro-optical functional components. The use of calcination heat treatment also allows the transformation of such structures from the initial polymer to final glass and glass-ceramic. Although the laser-induced damage threshold (LIDT) is an important parameter in characterizing all optics, it was not known for such sol–gel-derived glass microstructures. Here we present the first pilot study regarding this parameter, wherein functional microlenses have been made, damaged and calcinated for the series-on-one protocol. The results point to the fact that the LIDT can be increased significantly, even multiple times, thus expanding the usability of such resilient micro-optics.

Keywords: laser 3D printing; multi-photon lithography; glass micro-optics; light induced damage threshold; durable devices; photonic integrated circuits; high light intensity

Citation: Gailevicius, D.; Zvirblis, R.; Malinauskas, M. Resilient Calcination Transformed Micro-Optics. *Eng. Proc.* **2023**, *34*, 20. https://doi.org/10.3390/HMAM2-14270

Academic Editor: Kaupo Kukli

Published: 21 March 2023

1. Introduction

The field of laser multi-photon lithography is rapidly progressing. More and more interesting micro-optical devices are given form at the micro-scale. Examples include conventional [1] and Fresnel micro-lenses [2], holographic elements [3], meta-optics [4,5] multi-component systems [6], and most notably, micro objectives [7]. One problem overlooked with such systems is the laser induced damage threshold (LIDT) behavior. The major limitations stem from the fact that ease of manufacturing does not guarantee a high LIDT. Arguably not all applications are demanding in this regard [8], yet the domain of modern high-intensity pico- and femto-second pulses might be barred.

Some attempts to measure LIDT have been made and show variation depending on the following: if the resin used is organic or hybrid [9–11]; if a photo-initiator (PI) is used [12]; if the structure is a thin-film [11], bulk object [13] or a device [14,15]. Technically, the way to increase LIDT is by choosing resins without PI [16] and a less organic composition [17]. We want to go beyond this concept by using an alternative approach: making purely inorganic structures using a heat-based post-processing method called calcination, without disregarding the convenient multiphoton 3D printing method [18].

Essentially, for metal–organic systems such as SZ2080^{TM}[19] above 1000 °C [20], the resulting phase is an inorganic composite glass or glass-ceramic phase while retaining the printed geometry with homogeneous and repeatable shrinking [21]. Intuition dictates that transparent glassy [22] structures should feature higher LIDT values and, therefore, must be more resilient to high-intensity radiation, but this idea was never tested [14].

Therefore, the goal of this paper is to fabricate suspended and functional structures, Figure 1a, in this case, microlenses, heat-treat them, Figure 1b, and confirm the useful increase in LIDT, Figure 1c.

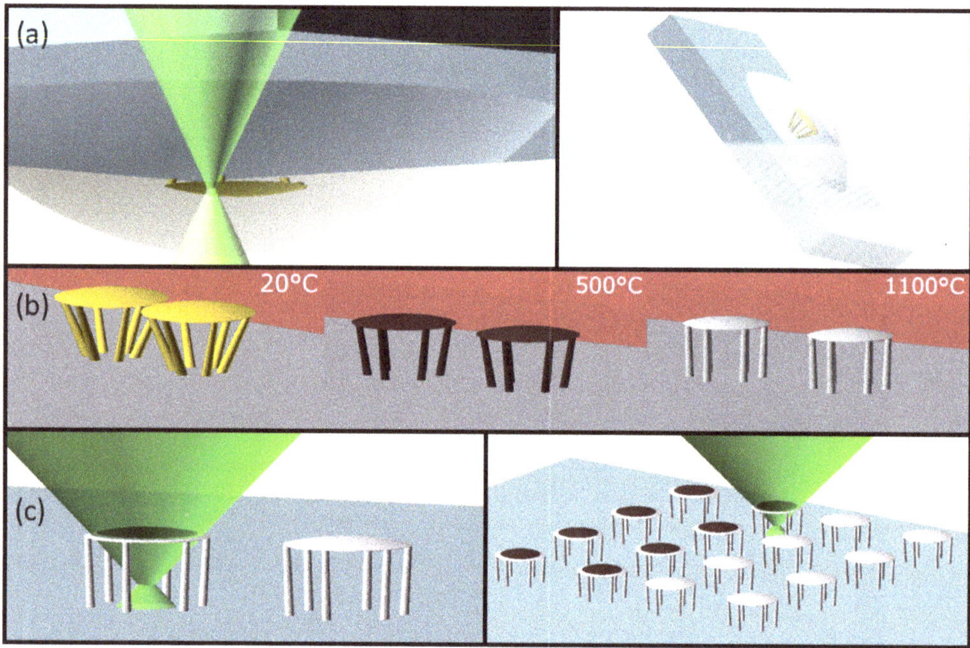

Figure 1. The experimental principle: (**a**) the fabrication of the test lens structure in resin, (**b**) heat treatment of the structure and compliant columns, (**c**) the exposure of the structures to measure the LIDT values.

2. Methods

To produce the lenses, we used the low-shrinkage organic–inorganic prepolymer SZ2080^{TM}. The preparation and exposure and conditions (Figure 1a) were selected similarly to Ref. [23] using a 1.4 NA objective, 517 ± 10 nm wavelength and 144 fs pulse duration and a repetition rate of 76 MHz. The lenses were printed on a quartz substrate. The final baseline geometry was of a plano-convex, 50 μm diameter, 300 μm focal length lens, with a thickness of approximately 2 μm. To support the lenses above the quartz substrate, pillars were printed with an inclination angle of 35° and a total height close to 30 μm. Heat treatment was performed at 1100 °C with a rise time of 12 h and held for 3 h. Some reactions with the ambient atmosphere are expected; however, the final transparent phase generated at the highest treatment temperature is of the essence here (Figure 1b). Non-calcinated (NCA) and calcinated (CA) samples were qualitatively examined and exposed to the probe beam in damage tests in an array form. (Figure 1c). Qualitative characterization of their imaging function was performed in a bright-field microscope to confirm their imaging function before and after LIDT measurements. See Figure 2a–d for the illustrated concept. The imaging function was used to confirm the occurrence of significant and catastrophic damage events (Figure 2b–d). After, they were characterized using a scanning electron microscope (SEM) and provided in Figure 3.

Damage tests were performed on arrays of micro-lenses. Arrays, mainly composed of 16 lenses, were divided in half to account for damage experiments for NCA (control) and CA (test) micro-optics. Damage tests were performed in the following sequence: an array of lenses is printed, half of the lenses are damaged before calcination, then the array is calcinated as described previously, and finally, the second half of the lenses are damaged. The laser system parameters used for all experiments were as follows: wavelength $\lambda_1 = 1030$ nm and $\lambda_2 = 515$ nm, repetition rate f = 200 kHz, pulse duration $\tau = 300$ fs, Plan-Apochromat Zeiss 20× objective (0.8 NA). S-on-1 damage tests were performed with both wavelengths, exposing the lenses for 50 ms and 5 s, corresponding to 10,000-on-1 and 1,000,000-on-1 pulses.

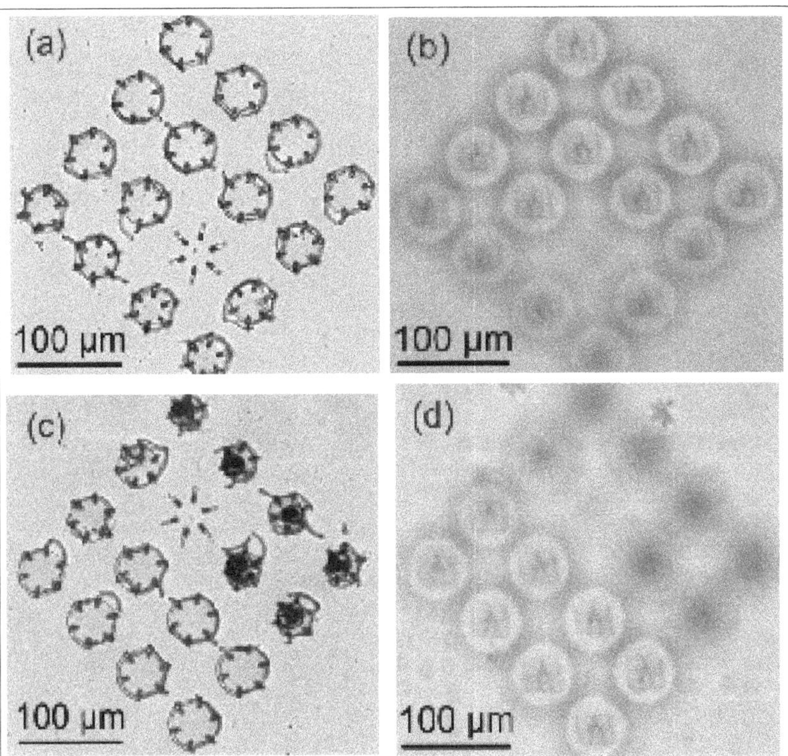

Figure 2. Optical characterization of micro-optics using an inverted microscope: (**a**) image of the surface of the optic components, (**b**) an image of the focal position (300 μm) of the same micro-optics, (**c**) bright-field image after damage, (**d**) degradation in image quality in the focal position of damaged lenses.

In addition, two damage protocols were tested. The first protocol is referred to as a local-damage protocol, where the beam diameter is approximately 4 μm ($1/e^2$ intensity level) on the sample. The second protocol is the delocalized-damage protocol, where the probe beam is 20 μm. The reasoning behind the delocalized damage protocol is to demonstrate the expected behavior where the full aperture of the lens is used.

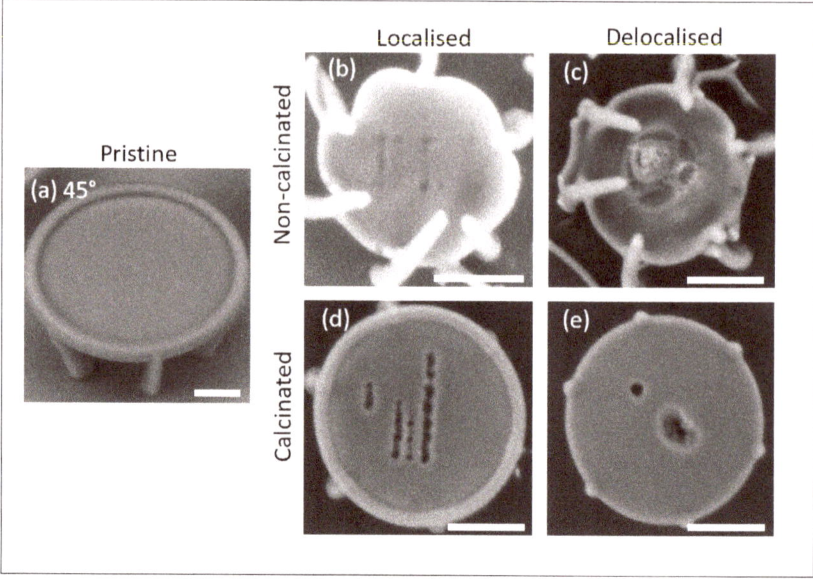

Figure 3. Scanning electron microscope micrographs of micro-lenses, illustrating the typical damage morphologies observed. The scale bar is 10 μm: (**a**) image of an NCA lens without damage, (**b**) local damage protocol lens damaged before CA, (**c**) delocalized damage protocol lens damaged before CA, (**d**) local protocol lens damaged after CA, (**e**) delocalized protocol lens damaged after CA. Tracks are formed for easier visualization of the damaged sites.

3. Results

3.1. Morphology

The observed damage morphology is shown in Figure 3. For comparison, an example of a pristine lens is given in Figure 3a. Such a lens initially is of 50 μm in diameter and, after calcination, shrinks down to 30 μm. The localized damage protocol produces small damage sites. The delocalized-damage protocol produces large damage sites for NCA lenses (Figure 3b). In this case, the lenses lose their imaging function as most of the aperture becomes distorted. The damage is technically catastrophic. On the other hand, the CA lenses shown in Figure 3d,e retain their qualitative imaging function and feature small diameter ablation sites reminiscent of fs-laser surface ablation. The morphology does not differ in any meaningful manner independent of the wavelength used. The only observed difference for NCA lenses is that brown discoloration is prominent, especially for $\lambda = 1030$ nm.

3.2. LIDT Values

The measurement results of LIDT values are summarized in Figure 4. We analyzed a combination of cases of localized and nonlocalized damage protocols, NCA and CA, 10^4-on-1 and 10^6-on-1, and 515 and 1030 nm wavelengths. The NCA localized damage results correspond well in the margin of error with the ones presented in the literature: $F_{1030} = 0.57$ J/cm^2 and $F_{515} = 0.13$ J/cm^2.

The non-localized damage thresholds do not correspond as accurately to previous results. They are lower. However, it is essential to note that the previously known experiment in which S-on-1 damage testing protocols were employed used only up to 1000 pulse exposure, so the current results with decreased LIDT are novel.

Damage thresholds of calcinated micro-optics from locally induced damage showed the highest increase—for 1030 nm damaging wavelength, a 3-fold increase was observed from $F = 0.6$–0.8 J/cm^2 to $F = 2.3$–2.7 J/cm^2, using a 300 fs laser pulse duration. This damage threshold was the highest out of all measured values and contained the entire exposure duration range from 50 ms to 5 s. The highest percentile increase in resilience was observed from using a second harmonic (515 nm)—around a 6-fold increase from uncalcinated micro-optics damage tests in all exposure durations from $F = 0.12$–0.17 J/cm^2 to $F = 0.8$–0.9 J/cm^2.

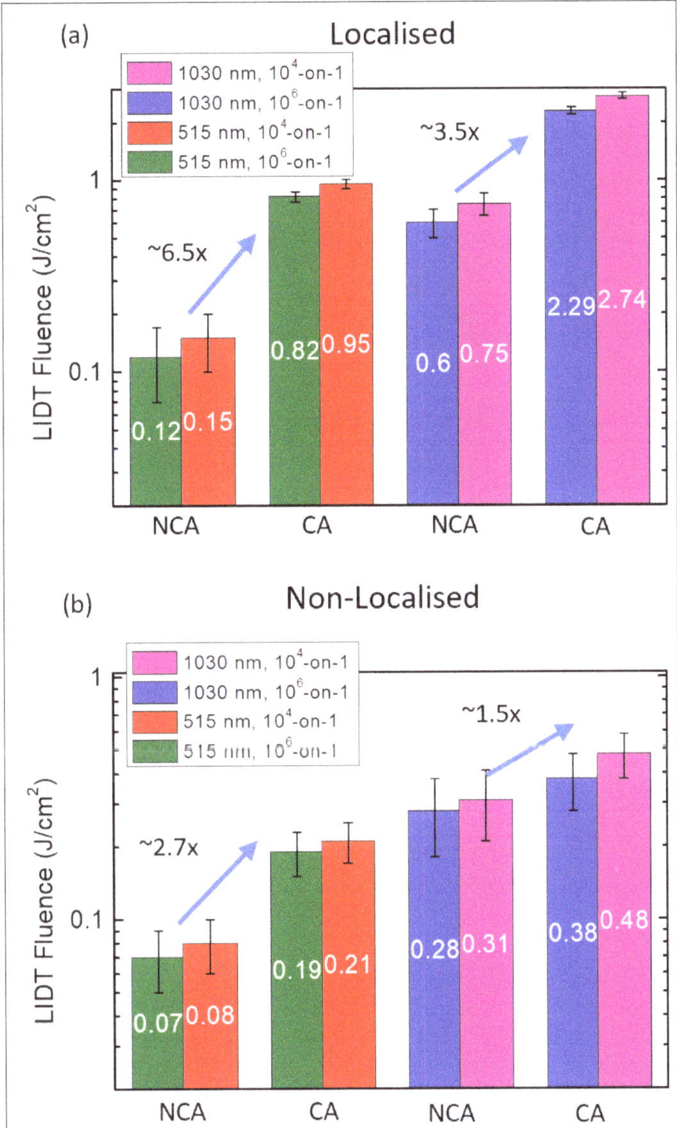

Figure 4. Light induced damage threshold measurement results for (**a**) localized damage protocol, (**b**) non-localized damage protocol.

A decrease in damage thresholds is also observed when the laser focus position is shifted, thus increasing the exposure area on the lens compared to uncalcinated counterparts. For the 1030 nm wavelength, the most minor calcination influence was measured—damage thresholds increased only approximately 0.5 times from $F = 0.28$–0.31 J/cm^2 for uncalcinated micro-optics to $F = 0.38$–0.47 J/cm^2 for calcinated lenses.

The main result observable for all cases is that the LIDT values always increase CA microlenses. The results are consistent, independent of irradiation area, testing process, etc.

4. Conclusions

Optical damage threshold measurements of SZ2080 material after calcination have been reported for the first time. The structures have been processed at a temperature of 1100 °C and feature a significant increase in damage threshold, reaching 300–600% as a conservative estimate. This increase supersedes all previously measured values [11,12,17]. In addition, the maximum measured LIDT value at λ = 1030 nm is F = 2.74 J/cm^2. It is relatively large as it approaches the level of fused silica [13] F = 3.11 J/cm^2. This result is promising, as applications of this material in harsh environments, as postulated many times previously, are proven for optical and IR wavelengths. Thus, the multiphoton lithography method combined with heat treatment can offer a technologically viable pathway to producing optical-grade, glass-level performance micro-optical elements.

Finally, as this is a pilot study, more research should be carried out for other significant regimes, such as nanosecond and continuous-wave damage tests. Regardless, the current results give credibility to the hypothesis that further research will also feature increased LIDT values.

Author Contributions: Conceptualization, D.G. and M.M.; methodology, D.G. and R.Z.; software, R.Z.; validation, R.Z., D.G. and M.M.; formal analysis, D.G.; investigation, R.Z.; resources, M.M.; data curation, D.G.; writing—original draft preparation, D.G.; writing—review and editing, M.M.; visualization, R.Z.; supervision, D.G.; project administration, M.M.; funding acquisition, M.M. All authors have read and agreed to the published version of the manuscript.

Funding: This research received funding from EU Horizon 2020, Research and Innovation program LASERLAB-EUROPE JRA Project No. 871124.

Institutional Review Board Statement: Not applicable.

Informed Consent Statement: Not applicable.

Data Availability Statement: Data available on request from the authors.

Acknowledgments: We acknowledge Maria Farsari and Vasileia Melissinaki for kindly providing the authors with the SZ2080^{TM} (IESL-FORTH, Heraklion, Greece) hybrid organic–inorganic materials for performing the described experiments.

Conflicts of Interest: The authors declare no conflict of interest.

References

1. Wu, D.; Chen, Q.D.; Niu, L.G.; Jiao, J.; Xia, H.; Song, J.F.; Sun, H.B. 100% Fill-Factor Aspheric Microlens Arrays (AMLA) With Sub-20-nm Precision. *IEEE Photonics Technol. Lett.* **2009**, *21*, 1535–1537. [CrossRef]
2. Asadollahbaik, A.; Thiele, S.; Weber, K.; Kumar, A.; Drozella, J.; Sterl, F.; Herkommer, A.M.; Giessen, H.; Fick, J. Highly Efficient Dual-Fiber Optical Trapping with 3D Printed Diffractive Fresnel Lenses. *ACS Photonics* **2020**, *7*, 88–97. [CrossRef]
3. Sandford O'Neill, J.; Salter, P.; Zhao, Z.; Chen, B.; Daginawalla, H.; Booth, M.J.; Elston, S.J.; Morris, S.M. 3D Switchable Diffractive Optical Elements Fabricated with Two-Photon Polymerization. *Adv. Opt. Mater.* **2022**, *10*, 2102446. [CrossRef]
4. Faniayeu, I.; Mizeikis, V. Realization of a helix-based perfect absorber for IR spectral range using the direct laser write technique. *Opt. Mater. Express* **2017**, *7*, 1453. [CrossRef]
5. Faniayeu, I.; Khakhomov, S.; Semchenko, I.; Mizeikis, V. Highly transparent twist polarizer metasurface. *Appl. Phys. Lett.* **2017**, *111*, 1–5. [CrossRef]
6. Žukauskas, A.; Malinauskas, M.; Brasselet, E. Monolithic generators of pseudo-nondiffracting optical vortex beams at the microscale. *Appl. Phys. Lett.* **2013**, *103*, 181122. [CrossRef]
7. Thiele, S.; Arzenbacher, K.; Gissibl, T.; Giessen, H.; Herkommer, A.M. 3D-printed eagle eye: Compound microlens system for foveated imaging. *Sci. Adv.* **2017**, *3*, 2. [CrossRef]
8. Sugioka, K.; Cheng, Y. Ultrafast lasers-reliable tools for advanced materials processing. *Light. Sci. Appl.* **2014**, *3*, 1–12. [CrossRef]
9. Stankova, N.; Atanasov, P.; Nikov, R.; Nikov, R.; Nedyalkov, N.; Stoyanchov, T.; Fukata, N.; Kolev, K.; Valova, E.; Georgieva, J.; et al. Optical properties of polydimethylsiloxane (PDMS) during nanosecond laser processing. *Appl. Surf. Sci.* **2016**, *374*, 96–103. [CrossRef]

10. Saha, S.K.; Divin, C.; Cuadra, J.A.; Panas, R.M. Effect of proximity of features on the damage threshold during submicron additive manufacturing via two-photon polymerization. *J. Micro Nano-Manuf.* **2017**, *5*, 031002. [CrossRef]
11. Žukauskas, A.; Batavičiūtė, G.; Ščiuka, M.; Jukna, T.; Melninkaitis, A.; Malinauskas, M. Characterization of photopolymers used in laser 3D micro/nanolithography by means of laser-induced damage threshold (LIDT). *Opt. Mater. Express* **2014**, *4*, 1601. [CrossRef]
12. Žukauskas, A.; Batavičiūtė, G.; Ščiuka, M.; Balevičius, Z.; Melninkaitis, A.; Malinauskas, M. Effect of the photoinitiator presence and exposure conditions on laser-induced damage threshold of ORMOSIL (SZ2080). *Opt. Mater.* **2015**, *39*, 224–231. [CrossRef]
13. Gallais, L.; Commandré, M. Laser-induced damage thresholds of bulk and coating optical materials at 1030 nm, 500 fs. *Appl. Opt.* **2014**, *53*, A186. [CrossRef] [PubMed]
14. Butkutė, A.; Čekanavičius, L.; Rimšelis, G.; Gailevičius, D.; Mizeikis, V.; Melninkaitis, A.; Baldacchini, T.; Jonušauskas, L.; Malinauskas, M. Optical damage thresholds of microstructures made by laser three-dimensional nanolithography. *Opt. Lett.* **2020**, *45*, 13–16. [CrossRef]
15. Simakov, E.; Gilbertson, R.; Herman, M.; Pilania, G.; Shchegolkov, D.; Walker, E.; England, R.; Wootton, K. Possibilities for Fabricating Polymer Dielectric Laser Accelerator Structures with Additive Manufacturing. In Proceedings of the IPAC 2018, Vancouver, BC, Canada, 29 April–4 May 2018; pp. 9–12. [CrossRef]
16. Samsonas, D.; Skliutas, E.; Ciburys, A.; Kontenis, L.; Gailevičius, D.; Berzinš, J.; Narbutis, D.; Jukna, V.; Vengris, M.; Juodkazis, S.; et al. 3D nanopolymerization and damage threshold dependence on laser wavelength and pulse duration. *Nanophotonics* **2023**, *12*, 1537–1548. [CrossRef]
17. Kabouraki, E.; Melissinaki, V.; Yadav, A.; Melninkaitis, A.; Tourlouki, K.; Tachtsidis, T.; Kehagias, N.; Barmparis, G.D.; Papazoglou, D.G.; Rafailov, E.; et al. High laser induced damage threshold photoresists for nano-imprint and 3D multi-photon lithography. *Nanophotonics* **2021**, *10*, 3759–3768. [CrossRef]
18. Merkininkaitė, G.; Aleksandravičius, E.; Malinauskas, M.; Gailevičius, D.; Šakirzanovas, S. Laser additive manufacturing of SiZrO$_2$ tunable crystalline phase 3D nanostructures. *Opto-Electr. Adv.* **2022**, *5*, 210077. [CrossRef]
19. Ovsianikov, A.; Viertl, J.; Chichkov, B.; Oubaha, M.; MacCraith, B.; Sakellari, I.; Giakoumaki, A.; Gray, D.; Vamvakaki, M.; Farsari, M.; et al. Ultra-Low Shrinkage Hybrid Photosensitive Material for Two-Photon Polymerization Microfabrication. *ACS Nano* **2008**, *2*, 2257–2262. [CrossRef]
20. Gailevičius, D.; Padolskytė, V.; Mikoliūnaitė, L.; Šakirzanovas, S.; Juodkazis, S.; Malinauskas, M. Additive-manufacturing of 3D glass-ceramics down to nanoscale resolution. *Nanoscale Horiz.* **2019**, *4*, 647–651. [CrossRef]
21. Gonzalez-Hernandez, D.; Varapnickas, S.; Merkininkaitė, G.; Čiburys, A.; Gailevičius, D.; Šakirzanovas, S.; Juodkazis, S.; Malinauskas, M. Laser 3D Printing of Inorganic Free-Form Micro-Optics. *Photonics* **2021**, *8*, 577. . [CrossRef]
22. Gallais, L.; Douti, D.B.; Commandré, M.; Batavičite, G.; Pupka, E.; Ščiuka, M.; Smalakys, L.; Sirutkaitis, V.; Melninkaitis, A. Wavelength dependence of femtosecond laser-induced damage threshold of optical materials. *J. Appl. Phys.* **2015**, *117*, 1–15. [CrossRef]
23. Butkus, A.; Skliutas, E.; Gailevičius, D.; Malinauskas, M. Femtosecond-laser direct writing 3D micro/nano-lithography using VIS-light oscillator. *J. Cent. South Univ.* **2022**, *29*, 3270–3276. [CrossRef]

engineering
proceedings

Proceeding Paper

Hologram Opens a New Learning Door for Surgical Residents—An Academic View Point [†]

Thivagar M

Department of Orthopaedics, General Hospital Jayanagar, Bangalore 560041, India; thivagarthedon@gmail.com
† Presented at the International Conference on "Holography Meets Advanced Manufacturing", Online, 20–22 February 2023.

Abstract: 3D images provide details of the human anatomy and activity of an internal organ of the body in high resolution. A 3D hologram is a highly efficient simulation technique, which could be effectively used for teaching and training for students in various aspects of the medical field at different levels.

Keywords: hologram; residents; HoloLens glasses

Citation: M, T. Hologram Opens a New Learning Door for Surgical Residents—An Academic View Point. *Eng. Proc.* **2023**, *34*, 21. https://doi.org/10.3390/HMAM2-14155

Academic Editor: Vijayakumar Anand

Published: 13 March 2023

1. Introduction

With the advent of newer technology and evolution of artificial intelligence, we can improve our expertise and decrease the errors in various medical fields. A hologram provides a non-contact three-dimensional (3D) image that can be seen with the naked eye. These 3D images provide details of the human anatomy and activity of an internal organ of the body in high resolution. Holography is a two-step process. In the first step, it records a hologram, in which a radiographic image is converted into a photographic record. The second step is to convert a hologram into a virtual image. Figure 1 [1].

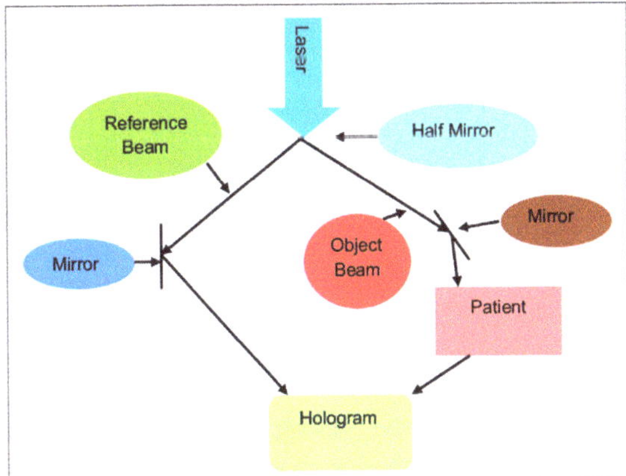

Figure 1. Process used in holography.

This provides better hands-on experience for medical and paramedical personnel. It can be employed in various fields of medicine, ranging from laboratory investigations to complex surgical procedures.

A 3D hologram is a highly efficient simulation technique, which could be effectively used for teaching and training for students in various aspects of the medical field at different levels.

Most residents specialize in a surgical field (especially in orthopedics) in a mid-level tertiary care center, peripheral center and post-graduate roles in corporate hospitals where the access to an anatomy dissection hall to learn in-depth anatomy would be difficult, which is required for better a understanding for the students while performing surgeries; to learn the techniques quickly, holograms will make it significantly easier for the students to achieve this. To err is human, but keeping our ethics in mind, it is permissible and acceptable in a simulated environment rather than on actual patients. Creating a simulated environment to learn reduces the damage to actual patients.

In this study, we focus on how holograms help in training orthopedic residents in various surgical techniques and critical situations to handle patients.

2. Methods

Due to constraints in the time period during residency in orthopedics, we are not able to gain adequate hands-on experience for most of the surgical procedures. The reasons are mainly due to the confidence of the consultants toward the resident, as we need to operate on actual patients and there should be minimal chances for mistakes. The role of the hologram is highlighted at this juncture. Using Microsoft's HoloLens glasses [2] in combination with the mixed reality system, it can project 3D holograms in real time under different clinical scenarios. Figure 2 shows that the process of obtaining surgical skills comes with practice, which, in turn, boosts the confidence level of the resident and the consultant, which is knowledge that cannot be gained from books alone. Hereby, holograms will provide different clinical, surgical and even emergency scenarios under simulation.

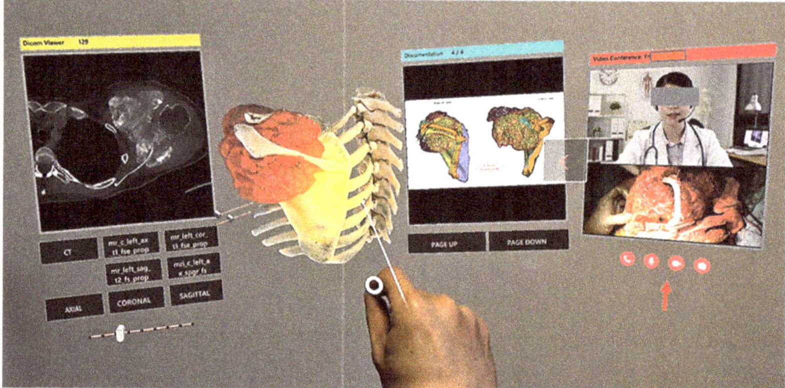

Figure 2. Hologram used for investigation and delineation of pathology.

The average total residency period for orthopedic surgeons is 2 or 3 years in India, which is less when compared to the courses abroad. The students have inadequate operating experience on actual patients, which may lead to a lack of confidence during the practice period. Using holograms, residents can have a real-time experience, which mimics a patient presenting with polytrauma along with an appropriate feedback system, which enables proper learning of the adequate resuscitative measures to help save the patient. This helps students deal with actual patients with a better level of confidence, and to minimize human errors. In India, due to the non-uniformity of exposure of orthopedic residents to various kinds of patients and surgical procedures, holograms help in bringing the learning exposure on a par. They can help the consultants sharpen their fine skills such as soft tissue procedures (tendon suturing, nerve grafting, etc.) with the appropriate simulation. Holograms can simulate a scenario and help in research that will be beneficial for patients. They help the consultant to superimpose the hologram image of the specific patients during surgery, such as spine in Figure 3, orthopedic oncology [3], etc., to predict the actual plane and degree of the implant and screw placement, and for proper resection of the tumor-free margin.

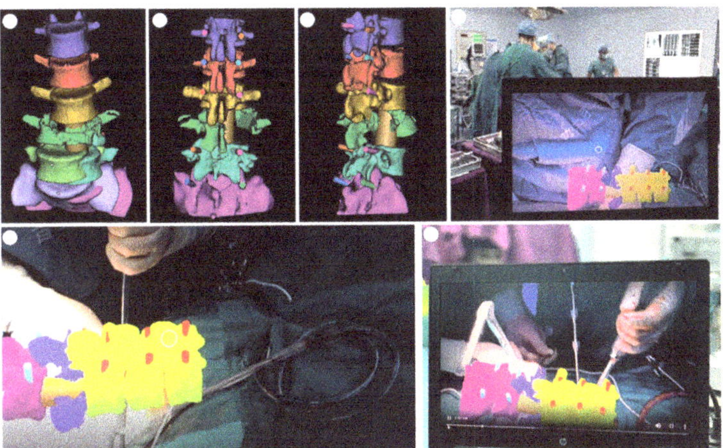

Figure 3. Hologram in spine surgery.

3. Results

This technology of using holograms that simulate various situations for surgeons and physicians may be helpful for better training residents. If this idea of using holograms for residents is beneficial, it can be extended even to undergraduates to develop knowledge about the basics in every specialty, which makes it easy to learn more things very early when they enter residency. It was found that surgeons who have started using holograms for a patient-tailored pre-op and intra-op planning had better outcomes. Hence, it will be more beneficial if this technology is extended for academic purposes.

4. Limitations

Considering extended reality will not be a better option for residents in certain scenarios, as it will give tactile feedback such as performing surgeries on actual patients, but it will enable them to learn better about anatomy and dissection, which is of the utmost importance for surgeons in their learning period.

In 1979, Piwernetz, et al. [4] published an idea of "Holography in orthopaedics" as a part of the "springer series in optical sciences" book, but it took many years to lead to a good knowledge base about this and create a high-resolution hologram for medical professionals, making this technology a difficult one to understand.

5. Conclusions

Holograms will change the learning technique of students by providing virtual imaging for better understanding the subject. Simulation of some basic surgeries will help residents to learn faster and decrease the incidence of errors in real-time surgeries in future. Holograms can change future methods of learning. However, there are certain limitations concerning hologram use for academic purposes, which includes the cost of manufacturing and maintenance and the source of funds to produce them. Hence, the feasibility of the advanced technology for residents is still debatable, for which monetary benefits are required from the government to every teaching institute.

Funding: This research received no external funding.

Institutional Review Board Statement: Not applicable.

Informed Consent Statement: Not applicable.

Data Availability Statement: Not applicable.

Conflicts of Interest: The author declares no conflict of interest.

References

1. Haleem, A.; Javaid, M.; Vaishya, R. Holography applications for orthopaedics. *Indian J. Radiol. Imaging* **2019**, *29*, 477–479. [CrossRef] [PubMed]
2. Lu, L.; Wang, H.; Liu, P.; Liu, R.; Zhang, J.; Xie, Y.; Liu, S.; Huo, T.; Xie, M.; Wu, X.; et al. Applications of mixed reality technology in orthopedics surgery: A pilot study. *Front. Bioeng. Biotechnol.* **2022**, *10*, 740507. [CrossRef] [PubMed]
3. Wong, K.C.; Sun, Y.E.; Kumta, S.M. Review and future/potential application of mixed reality technology in orthopaedic oncology. *Orthop. Res. Rev.* **2022**, *16*, 169–186. [CrossRef] [PubMed]
4. Piwernetz, K.; Von Bally, G. Holography in orthopedics. In *Holography in Medicine and Biology: Proceedings of the International Workshop, Münster, Fed. Rep. of Germany, 14–15 March 1979*; Springer: Berlin/Heidelberg, Germany, 1979; pp. 7–14.

engineering proceedings

MDPI

Proceeding Paper

Study of a pH-Sensitive Hologram for Biosensing Applications [†]

Komal Sharma [1,2], Heena [1], Girish C. Mohanta [1,2] and Raj Kumar [1,2,*]

1 CSIR-Central Scientific Instruments Organisation, Sector 30C, Chandigarh 160030, India; komalsharmasharma1998@gmail.com (K.S.); heenakhatri976@gmail.com (H.); gmohanta@csio.res.in (G.C.M.)
2 Academy of Scientific and Innovative Research (AcSIR), Ghaziabad 201002, India
* Correspondence: raj.optics@csio.res.in
† Presented at the International Conference on "Holography Meets Advanced Manufacturing", Online, 20–22 February 2023.

Abstract: Photopolymers are widely utilized as holographic recording media due to their ease of preparation and lack of wet chemistry post-processing. Holographic sensors constructed from a pH-sensitive photopolymer film have several applications in biosensors and the medical diagnostic field. However, the stability of photopolymer films in an aqueous medium is one of the most important challenges in their application for biosensing. Furthermore, the pH of the solution is another important parameter for biochemical reactions. In this work, we compared the pH sensitivity and stability of our holographic grating against two widely utilized classes of buffers; Phosphate Buffered Saline (PBS) and Tris-Acetate-EDTA (TAE) at two different pH values of 7.36 and pH 8.3, respectively. It was observed that a physiological pH (pH 7.4) had a negligible effect on the diffraction efficiency of the holographic sensor while it significantly deteriorated at a higher value of ~pH 8.3. This high sensitivity towards the minute pH difference of our holographic sensor could potentially be exploited for pH-based biosensing applications such as urea detection.

Keywords: holographic sensors; photopolymer film; biosensors; pH-sensing

1. Introduction

The quantitative and qualitative measurement of different physical and chemical processes is very important in the field of medical, agriculture, industrial and environmental applications. Holography is one of the optical techniques in which one can observe physical and chemical changes in the material. It is a method for storing and retrieving object information using light diffraction and interference. This technique uses an object beam and a reference beam that are both captured on the photopolymer film to produce an interference pattern. This technique has the advantage of not needing an additional rechargeable power unit, versatility, robustness and ease of preparation. These benefits led to its use in the field of biosensing applications such as glucose, lactose, pH detection and drug detection, among other things [1]. In order to achieve the necessary results with accuracy, selectivity, and sensing capacity, photopolymer films must be prepared with all points of care, mainly in biosensing applications.

Recently, it has been reported that the polyvinyl alcohol (PVA)-based photopolymer, due to its non-toxicity, swelling properties and good adhesive properties, have been used widely [2]. For biosensing applications such as drug detection and pH sensors, this binder is not desirable due to its hygroscopic nature. Additionally, cellulose acetate-based photopolymers are biocompatible, sustainable, cheap, flexible, lightweight and biodegradable [3] but have limitations in film formation due to the volatile nature of the solvent used.

In this paper, we propose to analyze the effect of pH change on the diffraction efficiency of transmission gratings recorded in a commercially available photopolymer as holographic

Citation: Sharma, K.; Heena, Mohanta, G.C.; Kumar, R. Study of a pH-Sensitive Hologram for Biosensing Applications. *Eng. Proc.* **2023**, *34*, 22. https://doi.org/10.3390/HMAM2-14272

Academic Editor: Vijayakumar Anand

Published: 21 March 2023

recording material by dipping it in the buffer solutions for different time intervals. The diffraction efficiency of the recorded grating can be measured as:

$$\text{D.E. (\%)} = \frac{I_{D+1}}{I_I} \times 100 \tag{1}$$

where, I_{D+1} and I_I are the intensity of the first-order diffracted beam and incident beam, respectively.

2. Materials and Methods

For biosensing applications, the sensitivity of the material toward pH plays a crucial role. In order to study its effect, changes in the diffraction efficiency of recorded holograms can be calculated at different pH exposure.

2.1. Recording Process

The transmission gratings were recorded in a two-beam holographic optical setup (Figure 1) using a continuous wave DPSS laser (532 nm). The beam was allowed to pass through a spatial filter (S.F.) followed by a lens (L) to obtain a collimated beam which was further divided into two by a beam splitter (B.S.). These two beams were directed by the mirrors M1 and M2 on the recording plate (photopolymer film, LLPF465, Light Logics, Trivandrum, India). This film requires a dosage of 12 mJ/cm^2 at 532 nm wavelength. The angle between the interfering beams was 25°. The intensity of these beams was adjusted to equal values by using a neutral density filter. A shutter was used to control the exposure from the laser, and the total recording intensity and exposure time were 12.8 mJ/cm^2 and 740 ms, respectively. After the recording process, gratings were characterized by a diffraction efficiency measurement (Figure 2) of the first diffracted order using an optical power meter. When the grating is illuminated with white light it diffracts different wavelengths at different angles resulting in a spectrum of colors as seen in Figure 2b.

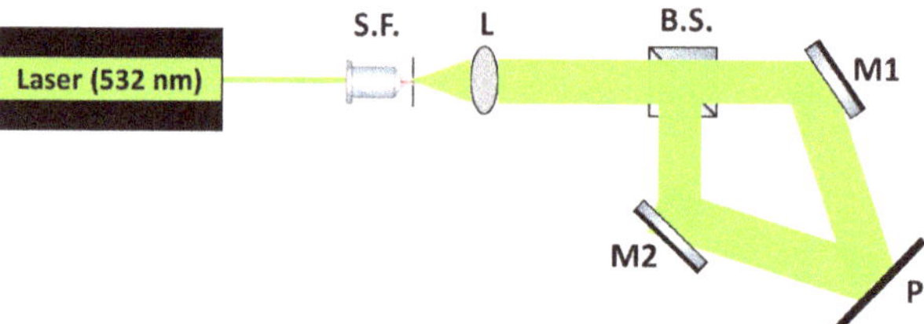

Figure 1. Schematic of experimental setup for recording of transmission gratings.

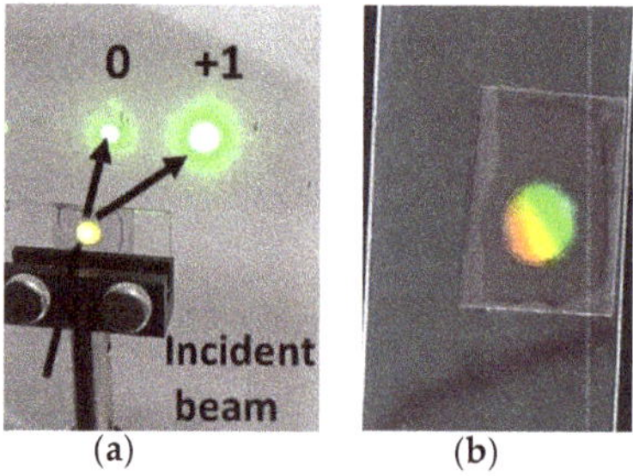

Figure 2. Photograph of (**a**) diffracted orders upon illumination with laser beam and (**b**) grating illuminated with in white light.

2.2. pH Sensitivity Measurement

In order to study the effect of pH sensitivity, three widely used buffer solutions were prepared with different pH values, namely Tris-Acetate-EDTA (pH-8.3) and Phosphate Buffered Saline (pH-7.4 and 8.3). The motive behind using PBS at two different pH levels was to examine the variation in the diffraction efficiency of a hologram with a varying pH of the same solution without relying on the pKa value. The pKa value was used to measure the acidity of a particular molecule in the solvent. Experiments were performed to see the effect on the photopolymer film by varying the pH when these gratings were dipped into different solutions for different time intervals, and a subsequent change in diffraction efficiency was measured. The total volume of the solution was 30 mL for the complete immersion of glass slides.

3. Results and Discussion

The recorded gratings exhibited a diffraction efficiency of >55% before any exposure to the pH solutions. Once these were allowed to completely dip into different pH solutions, a variation in the diffraction efficiency values was observed. The observed change was different for different pH values. For neutral pH, i.e., 7.4, the change in the diffraction efficiency was slight, within 5% of the original values (Figure 3a). On the other hand, when the pH increased to 8.3, the deterioration was much more for both PBS (Figure 3b) and TAE (Figure 3c) solutions, and the percentage change was up to 10% for the given dipping time intervals. From this, we could infer that the recorded holograms showed sensitivity toward alkaline pH and could be used for biosensing applications.

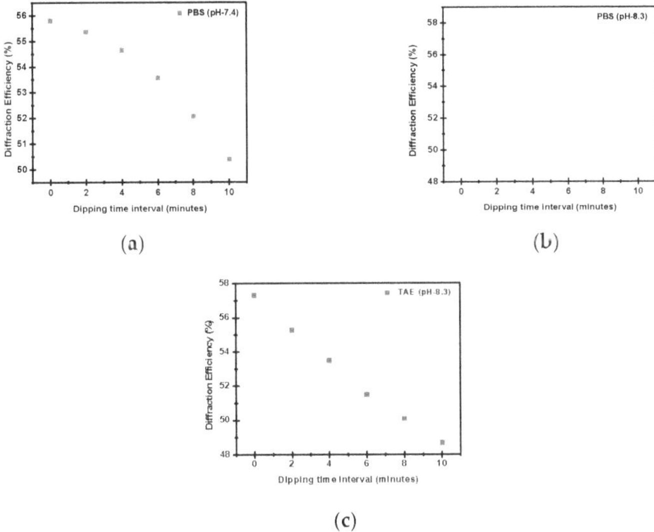

Figure 3. The diffraction efficiency response with dipping time intervals at different pH: (**a**) PBS pH-7.4; (**b**) PBS pH-8.3; (**c**) TAE pH-8.3.

4. Conclusions

In this work, we recorded holographic transmission gratings in photopolymer recording material and compared the pH sensitivity and stability of our holographic grating against two widely utilized classes of buffers; Phosphate Buffered Saline (PBS) and Tris-Acetate-EDTA (TAE) at two different pH values of 7.36 and pH 8.3, respectively. The sensitivity of the recorded hologram changed with pH, and by increasing its value from 7.4 to 8.3; the variance in diffraction efficiency decreased substantially. This study is helpful in the further development of holographic sensors for the bio-sensing industry, including glucose sensors, urea sensors, lactose sensors, and drug detection sensors.

Author Contributions: K.S.: methodology, formal analysis, writing—original draft preparation, H.: methodology, formal analysis, writing—original draft preparation, R.K.: Conceptualization, supervision, writing—review and editing, project administration, funding acquisition, G.C.M.: writing—co-supervision, review and editing. All authors have read and agreed to the published version of the manuscript.

Funding: This research was funded by the Department of Science and Technology, ministry of Science and Technology, India (DST—INSPIRE program, IF-210411) and CSIR funded project MLP-2014.

Institutional Review Board Statement: Not applicable.

Informed Consent Statement: Not applicable.

Data Availability Statement: The data presented in this study are available on request from the corresponding author. The data are not publicly available due to privacy and ethical reasons.

Conflicts of Interest: The authors declare no conflict of interest.

References

1. Yu, J.; Gai, Z.; Cheng, J.; Du, K.; Wei, W.; Li, Y.; Gao, Q.; Zou, C.; Qian, R.; Sun, Z.; et al. Construction of Beta-Cyclodextrin Modified Holographic Sensor for the Determination of Ibuprofen in Plasma and Urine. *SSRN Electron. J.* **2023**, *385*, 133650. [CrossRef]
2. Gaaz, T.; Sulong, A.; Akhtar, M.; Kadhum, A.; Mohamad, A.; Al-Amiery, A. Properties and Applications of Polyvinyl Alcohol, Halloysite Nanotubes and Their Nanocomposites. *Molecules* **2015**, *20*, 22833–22847. [CrossRef] [PubMed]
3. Gul, S.; Cassidy, J.; Naydenova, I. Water Resistant Cellulose Acetate Based Photopolymer for Recording of Volume Phase Holograms. *Photonics* **2021**, *8*, 329. [CrossRef]

Proceeding Paper

New Medical Imaging, Physics, Medical Need and Commercial Viability †

Zoltan Vilagosh

School of Health Sciences, Swinburne University of Technology, Hawthorn 3122, Australia;
zvilagosh@swin.edu.au

† Presented at the International Conference on "Holography Meets Advanced Manufacturing", Online, 20–22 February 2023.

Abstract: Successful medical diagnostic imaging tools satisfy three criteria; they produce useful information, they fulfill a diagnostic need and are financially viable. The present need is the development of more diagnostic modalities that display changes in the dynamic anatomy and function, "how things move or work, not just how things look". Bringing compartmentalized modalities together in a complimentary, real time, combined way could produce images with sensitivity and specificity that a single mode cannot offer alone. Conceivably, a combined magnetic resonance imaging/computerized tomography (MRI/CT) system with an added blend of modalities such as electrocardiographs, nerve conduction studies and electroencephalographs, all on the same platform, all performed simultaneously, with the images combined and analyzed together, could overcome the limitations of individual tests.

Keywords: medical diagnostics; combined diagnostic modalities; electromagnetic spectrum

1. Introduction

The proper implementation of medical practice needs information. In the diagnostic and management of patients, one starts with the history of the symptoms, followed by the physical examination. When the human senses are not sufficient, special tests come next. Medical diagnostic imaging is there to "see" when the examiner's senses are not enough. Any diagnostic imaging tool that comes into common use satisfies three criteria; it can be produced, it fulfills a diagnostic need, and it is financially viable. The "sweet spot" is where the physics, medical need and commercial viability intersect (Figure 1).

Citation: Vilagosh, Z. New Medical Imaging, Physics, Medical Need and Commercial Viability. *Eng. Proc.* **2023**, *34*, 23. https://doi.org/10.3390/HMAM2-14162

Academic Editor: Aravind Rajeswary

Published: 13 March 2023

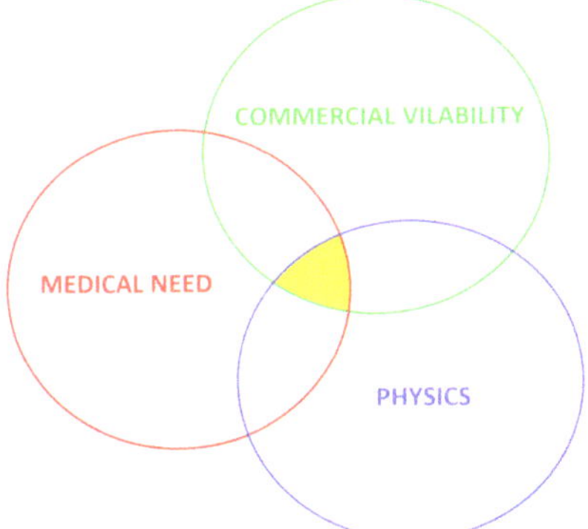

Figure 1. The intersection between the physics, medical need and commercial viability is the "sweet spot" where the opportunity medical imaging innovation exists.

2. The Physics

The dead hand of physics limits what is achievable. As Scotty from *Star Trek* was prone to say, "Ye cannae change the laws of physics"; medical diagnostic imaging tools are bound by what is physically possible. The realms of physics available for diagnostic use can be classified as electromagnetic waves, pressure waves and temperature. Humans are 60% to 65% water; thus, the physics of water becomes a dominant determinant. The absorption coefficient (α) of water, with respect to electromagnetic waves (in terms of cm^{-1}), is presented in Figure 2. The penetration depth (Dp), the depth at which about 37% of the radiation into water survives, is given by $Dp = 1/\alpha$. An α of 1.0 cm^{-1} gives a Dp of 1.0 cm, and for an α of 100 cm^{-1}, the Dp is 0.01 cm. For electromagnetic radiation to be useful as a deep tissue diagnostic modality, it needs to traverse the body, which means an α of 0.1 cm^{-1} or less.

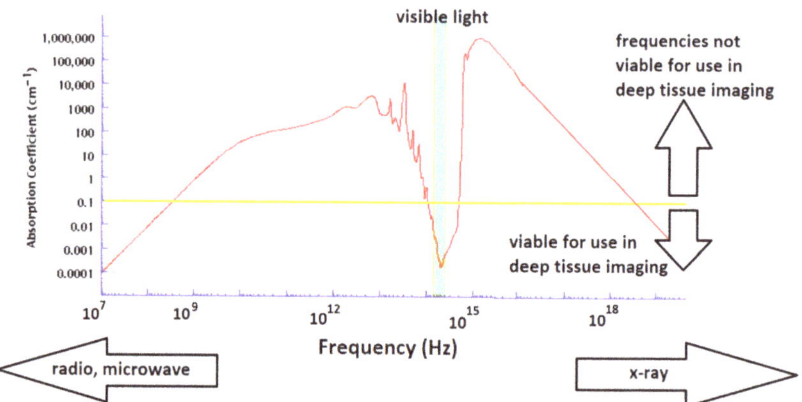

Figure 2. The absorption coefficient of water at room temperature. Data under 10^{16} Hz rely on [1]; data above 10^{16} Hz are extrapolated from [2].

Only radio waves, microwaves, visible light and X-rays have the desired penetration depth to capture the image of deep tissues (α in these regions is under 0.1 cm^{-1}, shown by the yellow line in Figure 2). Other electromagnetic frequencies may be useful for surface diagnostics. Radio waves and microwaves are utilized by electrocardiography, electroencephalography and magnetic resonance imaging (MRI), whist diagnostic X-rays, including computerized tomography (CT), use the X-ray spectrum. There are other constraints; these include the physics of the other chemical components in the body (which effectively rule out the visible spectrum for deep tissue studies), temperature limitations, radiation damage from X-rays, and even the psychological impact such as the panic caused by closed MRI in people who suffer from claustrophobia (%15 of the population). Image resolution, sensitivity (the capacity to resolve change) and specificity (the capacity to assign the change to a particular process) are also important considerations that are limited by physics. A similar analysis can be applied to pressure waves (sound and ultrasound). Sound attenuation is proportional to frequency (for tissue, the relationship ranges from linear to quadratic). Spatial resolution improves with higher frequency. This means there is a constant tradeoff between increasing the frequency to improve spatial resolution and lowering the frequency for better penetration depth in deep tissues. The usefulness of temperature differences is limited by the thermal diffusivity of the tissues.

3. Medical Need

The list of diagnostic imaging modalities that have found a lasting impact grows very slowly. Figure 3 is a photograph of a typical set of desk diagnostic tools in a medical doctor's office (it is missing a peak flow meter, the weight scales and height measurement device). The newest tool, the pulse oximeter—the small blue item—is based on visible and infrared light. It was invented in 1972 by Takuo Aoyagi. It only found common office use in the past decade. Other items are incremental improvements, such as a better and faster thermometer (to the right of the pulse oximeter) and the electronic blood pressure cuff.

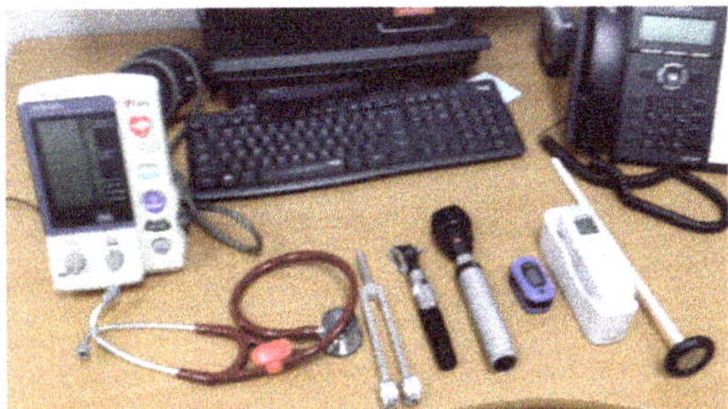

Figure 3. A typical medical office toolkit. From the left: the electronic sphygmomanometer, stethoscope, tuning fork, otoscope, ophthalmoscope, pulse oximeter, ear thermometer, tendon hammer.

The unmet medical need in an office diagnostic setting is a core hydration meter. Children with a fever and people of all ages who have gastroenteritis would benefit from an estimation of their hydration status. There are a number of unmet diagnostic imaging needs suited to a specialist imaging center setting. Versions of existing modalities that have improved sensitivity and specificity, do not involve the present level of risk and need less expertise or interpretation are on the list. There is also a need for better modalities that display the dynamic anatomy—"how things move or work, not just how things look". For example, there is a need to improve the contrast and accuracy of the capacity to detect meniscal cartilage lesions. One report puts the accuracy of MRI for knee cartilage lesions between 15% and 60% [3]. There are other gaps. Of the patients with a positive fecal occult blood (FOB) test, 58% will have normal colonoscopy, 39% will be diagnosed with a polyp and 3% will be diagnosed with a suspected cancer [4]. An anesthetic-free preliminary test that has better sensitivity than a FOB would be very welcome. It could reduce colonoscopies by 40% or more. Another example is the limitation put on diagnostic cardiac ultrasounds (echocardiography) due to the need for technicians with four years of training to perform the test. Cheaper, less "expertise intensive" imaging would enable greater access to these existing modalities that fill an existing need.

Some of the most difficult medical challenges revolve around episodic pain or other intermittent symptoms where no anatomical or biochemical anomalies can be found on any test. Diagnostic modalities that display changes in the dynamic anatomy in real time would be useful. Displaying changes in loose joints in the lumbar spine as they move or the motility of the stomach in response to a variety of foods would open new treatment options. Some attempts have been made in this direction, both with X-rays and MRIs. Both are limited by the field of view, and the X-ray modality is limited by the amount or radiation that can be used.

4. Commercial Viability

Any new device that has to be produced can needs to be justified in terms of its utility. An example of the failure of the financial criterion may be a plan for a new USD 1000 THz-based blood glucose monitor. There is definitely a clinical need for a blood glucose monitor, but that need has been filled. A complex finger prick-free continuous glucose monitor with mobile phone apps may cost $100. An example of the success of the commercial criterion is the MRI machine. It is complex, high maintenance and requires highly trained staff. Despite these factors, the medical need that MRI fulfills has made it a financially viable prospect.

5. Discussion, New Approaches and Conclusions

Physics is self-policing, meaning one cannot break its laws, but there are times when there are workarounds. Water absorbs GHz and THz frequency radiation too well to be useful in any non-surface imaging, but frozen water, ice, is 100 times more transparent at these frequencies. Freezing surface features or excised samples open the possibility of using GHz and THz frequency radiation as a diagnostic tool [5] for melanoma diagnosis or tissue ablation [6].

Ultrasound diagnostic devices are the most operator-dependent modalities. They can take a long time to acquire images and need trained people for both image acquisition and interpretation. Reducing the human factor via automation, the use of artificial intelligence and machine learning, would improve access and reduce cost.

Medical diagnostics is currently segregated. One uses either an MRI, a CT or an ultrasound. Often, a diagnostic report from one modality suggests further tests in another. It is time to bring the modalities together in a complimentary, combined way, with images fused to show sensitivity and specificity that neither mode can offer alone, in a faster time frame. Perhaps a combined MRI/CT system on the same platform, with the images combined and analyzed together, can be a solution. The idea is not novel [7], but as the unit cost of MRI machines fall, it may be the time for it to become a reality. Passive tests such as electrocardiographs and electroencephalographs could also be considered for inclusion in these combined modalities.

The best prospects for commercial success may be techniques that replace a procedural intervention with an imaging modality. The challenge with those is not to lose sensitivity or specificity.

Funding: This research received no external funding.

Institutional Review Board Statement: Not applicable.

Informed Consent Statement: Not applicable.

Data Availability Statement: No new data were created.

Conflicts of Interest: The authors declare no conflict of interest.

References

1. Segelstein, D.J. The Complex Refractive Index of Water. Ph.D. Thesis, University of Missouri, Kansas City, MI, USA, 1981.
2. Hill, R.; Holloway, L.; Baldock, C. A dosimetric evaluation of water equivalent phantoms for kilovoltage x-ray beams. *Phys. Med. Biol.* **2005**, *50*, N331. [CrossRef] [PubMed]
3. Koch, J.; Ben-Elyahu, E.J.R.; Khateeb, B.; Ringart, M.; Nyska, M.; Ohana, N.; Mann, G.; Hetsroni, I. Accuracy measures of 1.5-tesla MRI for the diagnosis of ACL, meniscus and articular knee cartilage damage and characteristics of false negative lesions: A level III prognostic study. *BMC Musculoskelet. Disord.* **2021**, *22*, 124. [CrossRef] [PubMed]
4. Parkin, C.J.; Bell, S.W.; Mirbagheri, N. Colorectal cancer screening in Australia. *Aust. J. Gen. Pract.* **2018**, *47*, 859–863. [PubMed]
5. Vilagosh, Z.; Lajevardipour, A.; Wood, A.W. Computational phantom study of frozen melanoma imaging at 0.45 terahertz. *Bioelectromagnetics* **2019**, *40*, 118–127. [CrossRef] [PubMed]
6. Vilagosh, Z.; Lajevardipour, A.; Wood, A.W. Imaging and lesion ablation modeling in skin using freezing to enhance penetration depth of terahertz radiation. In Proceedings of the Photonics in Dermatology and Plastic Surgery, San Francisco, CA, USA, 2–3 February 2019; Volume 10851, pp. 29–37.
7. Wang, G.; Feng Liu, F.; Liu, F.; Cao, G.; Gao, H.; Vannier, M.F. Design proposed for a combined MRI/computed-tomography scanner. *SPIE Newsroom*, 11 June 2013. [CrossRef]

Proceeding Paper

An Efficient Designing of IIR Filter for ECG Signal Classification Using MATLAB †

Nandi Manjula [1], Ngangbam Phalguni Singh [1,*] and P. Ashok Babu [2]

[1] ECE Department, Koneru Lakshmaiah Education Foundation (K L Deemed to be University), Vaddeswaram 522302, Andhra Pradesh, India; manju.reddy474@gmail.com

[2] ECE Department, Institute of Aeronautical Engineering, Hyderabad 500043, Telangana, India; p.ashokbabu@iare.ac.in

* Correspondence: phalsingh@gmail.com

† Presented at the International Conference on "Holography Meets Advanced Manufacturing", Online, 20–22 February 2023.

Abstract: The electrocardiogram (ECG) is a biological signal that is frequently employed and plays a significant role in cardiac analysis. In the analysis of important indicators of the distribution of patients' ECG record, the R wave is crucial for both analyzing abnormalities in cardiac rhythm and determining heart rate variability (HRV). In this article, a brand-new method for classifying and detecting QRS peaks in ECG data based on artificial intelligence is provided. The integration of the ECG signal data is proposed using a reduced-order IIR filter design. To construct the reduced-order filter, the filter coefficient using the min–max method. The main focus of this study is on removing baseline uncertainty and power line interferences from the ECG signal. According to the results, the accuracy increased by about 13.5% in comparison to the fundamental Pan–Tompkins approach and by about 8.1% in comparison to the current IIR-filter-based categorization rules.

Keywords: ECG; interpretation; acquisition; HRV; Pan-Tompkins method; min–max method

1. Introduction

The World Health Organization has determined that heart arrest is the leading cause of mortality worldwide. Due to the strong emphasis on medicine, preventive measures, and technology in cardiac health research, investigators have been working to develop the cardiovascular abilities that are typically used in clinics [1,2]. The majority of heart pathology can be understood by looking at the ECG signal. Heart rate and ECG signals are used to evaluate a healthy heart. A cardiac arrhythmia is recognized if an ECG is recorded from a patient and there is any nonlinearity. Figure 1 depicts an average ECG rhythm. The PQRS-TU wave's length and amplitude provide important information regarding the severity of the heart disease. In the clinical setting, the ECG signal is subjected to a variety of sounds during acquisition [3]. Important cardiac foundations include frequency determination, signal superiority, noise, and power line interference (PLI), in addition to external electromagnetic field intrusion [4,5]. It is advised that the issue of contaminated noise removal be solved because it improves accuracy and is crucial for the ECG data. Medical professionals use ECG extensively in the assessment and identification of cardiac health. Cardiologists frequently use ECG as a diagnostic tool to identify cardiovascular diseases [6–8]. The early stages of heart disease are crucial because they can lessen abrupt cardiac failure. Accurate diagnoses require ECG signals of good quality. Electrodes are used to display and record electrical cardiac activity on the body's skin [9].

Normal sinus rhythm (NSR), often known as a regular heartbeat, is present in healthy hearts [10–12]. Atrial depolarization is indicated by P-waves. Regular Q waves indicate septal depolarization and are an early descending deflection of the P wave. The ECG's most common waveform for identifying and detecting early ventricular depolarization is the R

Citation: Manjula, N.; Singh, N.P.; Babu, P.A. An Efficient Designing of IIR Filter for ECG Signal Classification Using MATLAB. *Eng. Proc.* **2023**, *34*, 24. https://doi.org/10.3390/HMAM2-14154

Academic Editor: Vijayakumar Anand

Published: 13 March 2023

wave. which shows the late ventricular depolarization [13,14]. Ventricular repolarization is characterized by the T-wave. The Purkinje fibers that show the most recent ventricular residuals, or U waves, repolarize.

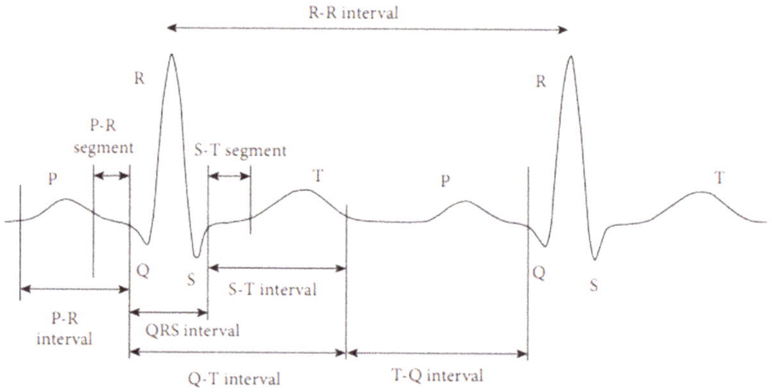

Figure 1. Two-cycle regular ECG waveform.

2. Relevant Review Work

As communication technology has developed, wireless communication has become one of the most popular ways for people to conveniently exchange ideas and thoughts. An audio noise-reduction system is a technique for eliminating noise from audio transmissions. Devices for audio noise reduction use two main strategies. The complementary type calls for carefully removing the audio signal before recording (primarily on tape) [1–5]. The relative weight that the filter assigns to data samples is determined by these settings. Typically, to accomplish this, particular frequencies or frequency ranges must be eliminated [6–13]. Contrarily, filters have a wide range of extra aims, particularly in the area of image processing. They do not merely operate in the frequency domain. There is no apparent hierarchy, but there are many classification systems, all of which overlap in different ways. Filters come in both linear and nonlinear varieties. The system's properties include shift invariance, which is sometimes known as time-variant or time-invariant [14–17]. A filter that operates in a spatial domain is said to be space invariant. The filters that handle time- or spatial-domain signals are slow in time [18,19]. Unlike continuous-time filters, which can be both active and inactive, discrete-time filters can only be active or inactive. This study illustrated the effective use of an optimization-based filter for identifying and categorizing ECG peaks. It participated in both passes. In the first pass, a useful lower-order IIR filter design method based on transfer function optimization was suggested for detecting QRS peaks. For peak detection performance improvement, the IIR filter and Hilbert transform was used. Three distinct approaches were tested for their filtering effectiveness for baseline wondering.

The majority of people today aged between the ages of 40 and 60 experience cardiac-related health problems. An electrocardiogram is the best way to capture heart impulses and identify any abnormalities at an early stage, as demonstrated in Figure 2a, showing anormal ECG, and Figure 2b, showingan abnormal ECG. The heart rate variability and QRS complex are used to classify the abnormalities [20]. For the purpose of removing motion distortions from the ECG data, numerous authors have put forth various methods. The aforementioned techniques include wavelet transforms (WTs), adaptive filters (AFs), and empirical mode decomposition (EMD) [21]. Blanco-Velasco proposed an ECG enhancement technique to remove baseline drift and noise brought on by high frequencies. On the other hand, mode mixing, which yields erroneous intrinsic mode functions, is one of the prevalent problems in empirical mode decomposition.

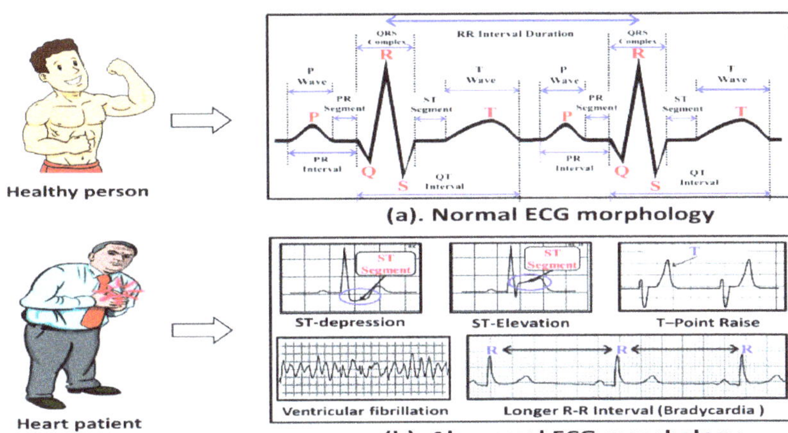

Figure 2. (**a**) Healthy ECG; (**b**) diseased ECG.

3. Objectives of Research

Designing an efficient IIR filter for ECG signals to identify heart problems was the main focus of the design, analysis, and implementation of the following subsystems. They are categorized as aspects of the work flow in order to create an effective model for ECG analysis.

- The first theory is furthered by the implementation of modules such arithmetic circuits for filters used in ECG signal classification, as well as the construction of parallel prefix circuits with the least amount of depth utilizing the FPGA hardware prototype.
- The second theory is that by creating an algorithm that can identify the difficult QRS problem in real-time ECG classification, we may further investigate the effective filter utilized in ECG signal classification.

4. Pan–Tompkins Peak Detection Approach

There have been numerous strategies created to enhance the effectiveness peak detection and classification efficiency, which are anticipated to be directly correlated with the effectiveness of the methods used during the preprocessing step. Thus, two current filtering techniques—Pan–Tompkins and a 60-order IIR filter, respectively—were used. For peak identification, the Pan–Tompkins technique was most frequently utilized. However, other variations of the Pan–Tompkins approach were developed to increase the categorization effectiveness of the ECG signal using the updated peak detection algorithm.

5. IIR Filter Design

It is recommended to use the optimization method in order to bring down the expected order of the IIR filter design. A block diagram of the proposed ECG categorization algorithm can be found in Figure 3a. In the proposed IIR filter, which is a two-stage filter, the pass band filters and stop band filters are laid out in the manner that is depicted in Figure 3b.

$$Y(n) = X(n) * H1 * H2, \tag{1}$$

where H1 is the pass band filter's transfer function; and an example of a transfer function for 106 MIT-ECG BIH's data is provided as

$$H1 = \frac{0.20346s^4 - 0.7131s^2 + 0.24566}{s^4 + 0.5488s^3 + 0.4535s^2 + 0.1763s + 0.1958} \tag{2}$$

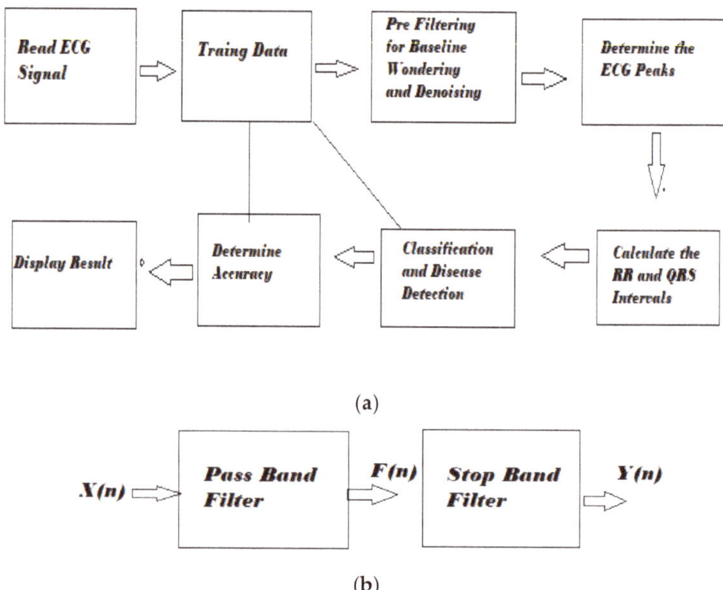

Figure 3. (**a**): Image detection and classification methods.(**b**): Basic IIR filter design procedure in two stages.

The optimization method is suggested to reduce the expected IIR filter design's order:

$$H2 = \frac{\begin{array}{l}0.3201s^{16} + 4.517s^{15} + 30.45s^{14} + 130s^{13} + 393.2s^{12} + 892.9s^{11} + 1574s^{10} + 2196s^9 \\ + 2452s^8 + 2196s^7 + 1574s^6 + 892.9s^5 + 393.2s^4 + 130s^3 + 30.45s^2 + 4.517s + 0.3201\end{array}}{\begin{array}{l}s^{16} + 12.12s^{15} + 70.25s^{14} + 258.1s^{13} + 672.2s^{12} + 1316s^{11} + 2004s^{10} + 2419s^9 \\ + 2340s^8 + 1819s^7 + 113s^6 + 559.4s^5 + 214.8s^4 + 62s^3 + 12.7s^2 + 1.649s + 0.102\end{array}} \tag{3}$$

Equations (1) and (2) depict the transfer function for the IIR filter, the pass band filter, and the stop band filter, respectively (3).

5.1. Optimization Methods

The methods of ECG processing that are used most frequently are discussed below.

5.1.1. The Low-Pass Differentiation Approach (LPD)

The total band-pass filter significantly reduces the effects of all sorts of unwanted interference and frequency noise. This algorithm chooses the necessary QRS complexes based on a set of thresholds and modifies the threshold according to the magnitude of the peak. Thethresholds for this algorithm are determined by a human component. The operator's threshold-setting expertise may be the main source of error in this Pan–Tompkins method.

5.1.2. Hilbert Transform (HT)

HT use the 90° phase angle shift method of each component of the signal h(t). The Hilbert transform of h(t), using h(t) as its signal representation, can be written as

$$\hat{h} = \frac{1}{\pi} \int_{-\infty}^{\infty} \frac{h(p)}{t - p} dp \tag{4}$$

The HT approach was employed by researchers in Refs. [13,21–23] implemented for the recognition of ECG QRS.

5.1.3. Optimum Reduced-Order IIR Filter Design

The basic IIR filter requires 16 orders, as can be seen from the equations above. In this study, MINMAX optimization is proposed as a way to reduce the order of the transfer function. Figure 3 shows the image detection and classification methods in sequence.

$$\min x1 \in X \max x2 \in X f(x1, x2) \tag{5}$$

where X Rdx1 and Y Rdx2 are convex sets, f is a differentiable objective function, and x1 and x2 are optimization variables. Transfer function coefficients are optimized in a sequential manner, as demonstrated in the sequential min–max optimization technique.

1.   To assess a particular instance, a sample was suggested;
2.   It was suggested that an optimization approach be designed for the denoising of the herring assistance signal using a minimized-order IIR filter;
3.   Two different filters must consider the proposed IIR filter before denoising.

6. IIR Filter Design Algorithm

The following list describes each step taken in the suggested algorithm:

i.   The ECG data containing occurrences of arrhythmia has 48 channels;
ii.   ECG characteristics are defined;
iii.   The standard QRS interval (0.098 s) and sampling frequency Fs (500 Hz), i.e., the QRS (t), are established;
iv.   A 60-order IIR high pass filter is used to perform baseline wondering;
v.   The best 150 Hz IIR low-pass filter possible is created. An upper and lower cutoff frequency FL and FH are designed for a pass band Butterworth filter. A stop band is created (Figure 3b). FL1 and FH1 are the lower and higher cutoff frequencies for the Butterworth IIR filter;
vi.   With a 100-coefficient IIR stop band filter, interference from power lines is eliminated;
vii.   The average of the regular and irregular heart rates is determined.

The 30 ECG data, which contain a wide variety of ECG data, were chosen for the current investigation out of the 48 channels that humans have available. Figure 4 displays the input ECG data. At a rate of 360 samples per second, the following ECG data were captured. The ECG information is displayed in Figure 5a–f. The ECG data show the most erratic variations in these channels. Therefore, it is important to show how well the peak detection method works against them. These channels were chosen for another reason; several ECG peak detection techniques already in use, such the ones in Refs. [24–34], also take them into account.

6.1. Min–Max Optimization Enabled Filter Design

This research suggests a fundamental change to the ECG signal processing procedure in order to develop an optimal reduced-order IIR filter. In order to achieve smoothing, the proposed lower-order IIR filter design was used in place of the traditional FIR low-pass filters. The sequential results for the proposed IIR filter using optimization strategies for filtering the ECG signal are shown in the last row in results, making it evident that the projected technique significantly smooths out the artifacts while also maintaining the characteristics of the QRS peaks.

6.2. Results of Optimized Filter ECG Design

Figure 4 provides an outcome comparison for the suggested IIR filter design. Three filters are given after the results with a stop band as an IIR filter.

6.3. Simulation Results of QRS Peak Detection in MATLAB

From Figure 5 the proposed filter design, as can be seen, increases the magnitude of the following data, which are included in this collection: 100, 101, 102, 103, 104, 105, 100,

101, 102, 103, 104, 105, 106, 107, 108, 109, 111, 113, 114, 200, 201, 202, 203, 207, 208, 209, 210, 212, 213, 214, 215, 217, 219, 221, 222, and 228.

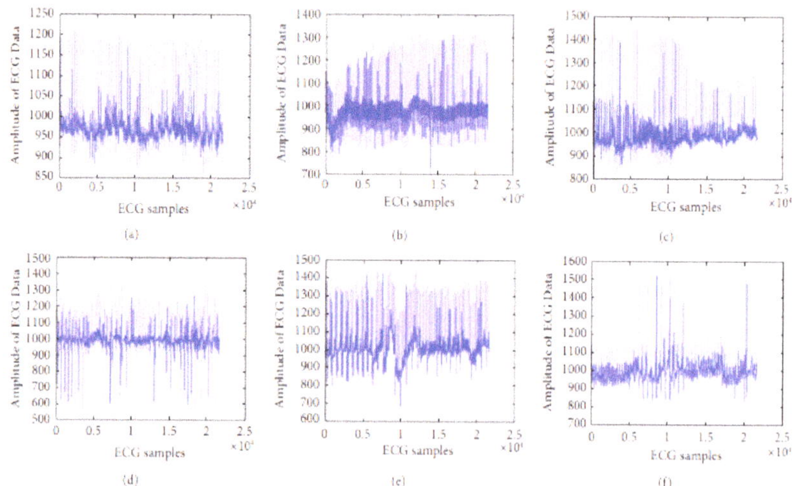

Figure 4. Data from the 5.556-s ECG arrhythmia database of 35 people recorded for more than 25 min at the MIT-BIH/Physio Net were used in the current investigation.

Figure 5. For perceptual result representation, six distinct ECG data with various feature difficulties were taken into account: (**a**) 102, (**b**) 106, (**c**) 109, (**d**) 200, (**e**) 208, and (**f**) 228.

6.4. ECG Image Classification

Three distinct ECG classification rules were offered in our article. In the guidelines for the HRV-based ECG classifications of regular or irregular ECGs are enumerated. When the method fails to locate a precise beat, a false negative (Fn) is produced. Fns are taken out of the corresponding annotation case in the MIT-BIH record. An inaccurate beat outcome is referred to as a false positive (Fp), whereas a true positive (Tp) is indicated by a precise beat recognized using the suggested approach. Additionally, a true negative (Tn) is accurate, and does not include detected beats. From Figures 6 and 7 we can analyze the ECG signals of the optimized and Hilbert transform filtered signals with respect to R peak detection efficiency.

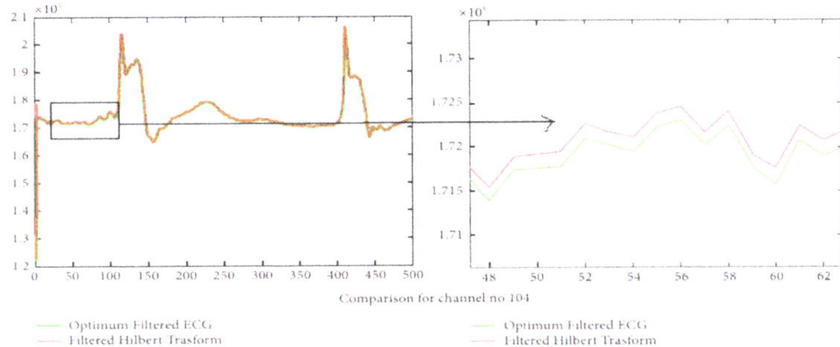

Figure 6. Hilbert transform for channel number 104.

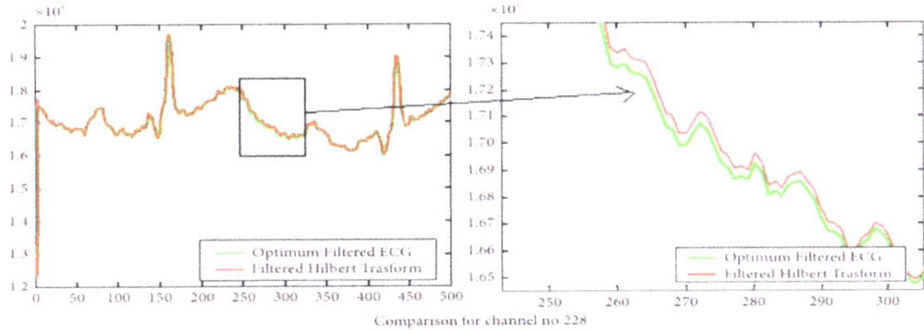

Figure 7. The channel number 228's R peak detection efficiency.

The original ECG signal's characteristics are preserved as well. With the improved filtering representation for the 3500 and 2000 initial samples, which is clearly displayed in Figure 8, further analysis becomes much clearer. The results for the ECG 110 m signal are shown in Figure 8a. Figure 8b shows a zoomed-in view of the filtering, which makes the baseline filtering effects for 2000 samples very obvious.

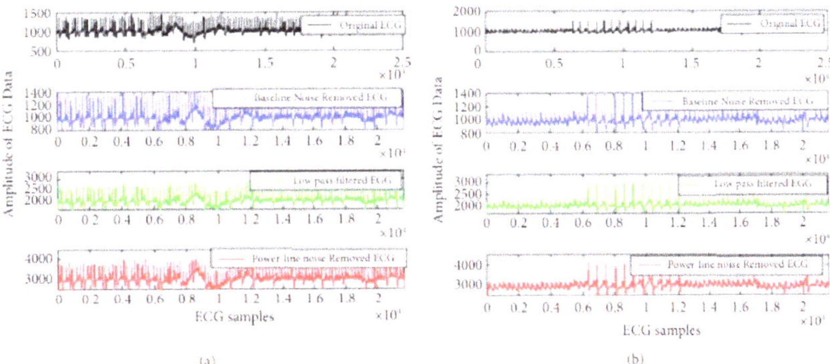

Figure 8. For the 3500 samples, ECG signal pre-processing is plotted for the (**a**) 101st ECG data and (**b**) 106th ECG data.

Figure 8a,b shows the number of samples taken into consideration for the experiment on the *x*-axis.

6.5. Time Domain HRV Parameter Analysis

The proposed ECG classification and peak detection approach includes the examination of time domain statistical HRV characteristics. Using the same RR intervals that were used to analyze the ECG signal, several parameters in the time domain were determined. The following definitions apply to the parameters which this study employed for analysis. Table 1 displays the statistical results of the measured parameters for the six input ECG signals using the suggested peak detection method. According to Table 1, SDNN was produced at its highest level for 228 m channels and at its lowest level for 106 m ECG channels.

Table 1. Comparative analysis for statistics parameters of existing method with proposed method.

Ref. No.	Accuracy	QRS Peak Detection	RR Interval	HRV Analysis
[2]	85%	No	No detected	No
[4]	85.6%	No	No detected	No
[6]	88%	No	No detected	YES
[8]	89%	No	No detected	No
[14]	90%	YES	detected	YES
[20]	91%	YES	detected	No
[26]	92.67%	YES	detected	YES
This work	96.87%	YES	detected	YES

Maximum and lowest heart rate evaluations were conducted for the 106 m ECG channels, respectively. The most effective method for keeping track of the dynamic shift in self-care under anesthesia may be Poincare plotting. The point on the plot is represented by the value of each subsequent pair of RR intervals.

From Figure 9 The maximum evaluation of the parameter The minimal RMSSD was 106 m ECG channel for the following maths problems: 102m.mat, 106m.mat, and 102m.ECG. The following definitions apply to the parameters this study employed for analysis:

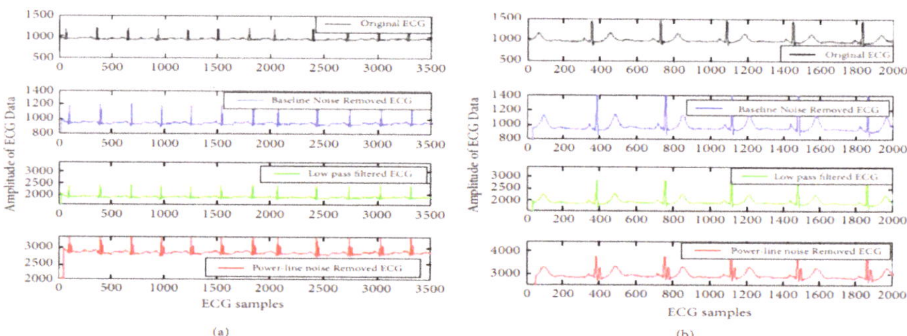

Figure 9. A signal provides a effects of ECG signal artifact reduction. (**a**) 101st ECG data, (**b**) 106th ECG data.

From Figure 10a–d, we are analyzing QRS peaks in signal 102 m, 109 m, 208 m and 228 m.

(a) Standard deviation of NN interval (SDNN): For each pair of RR intervals is used to define the SDNN.

(b) Root mean square SD (RMSSD):

$$RMSSD = \sqrt{1/M\left(\text{diff}(RR\ \text{region})^2\right)} \tag{6}$$

where M is the RR interval vector's length, denoted by the symbol RR region.

(c) NN50 value: The number of R to R intervals that are longer than the 50 ms interval is known as the NN50 value.

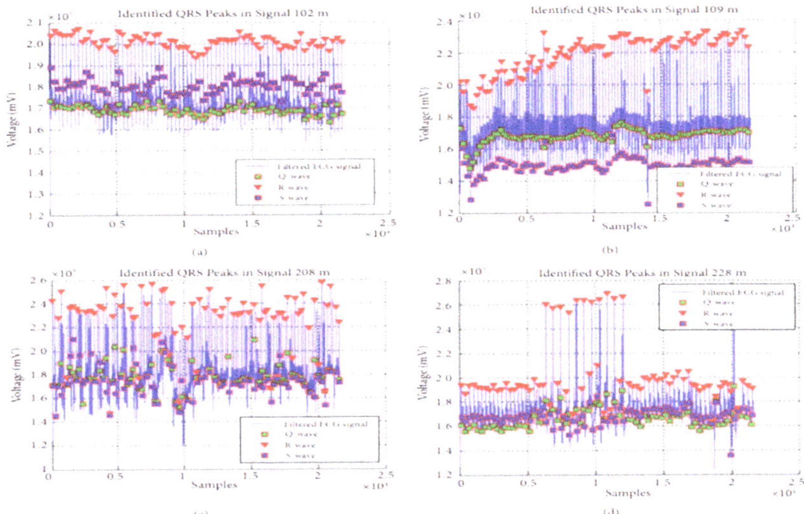

Figure 10. Results of the suggested optimum IIR filter technique for QRS peak identification for four ECG signals.

The sequential results for the proposed IIR filter using optimization strategies for filtering the ECG signal are shown in the last row in Figure 10, making it evident that the projected technique significantly smooths out the artifacts while also maintaining the characteristics of the QRS peaks. Figure 11a,b are representing the signal preprocessing of the data for 101st and 106th ECG samples respectively.

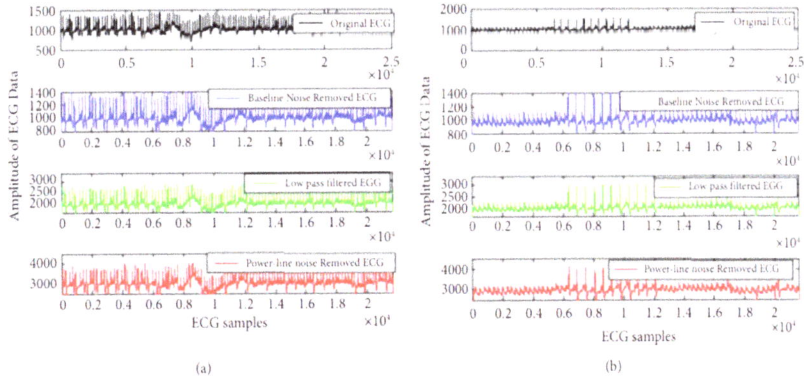

Figure 11. For collected more samples, ECG signal pre-processing is plotted for the (**a**) 101st ECG data and (**b**) 106th ECG data.

By analyzing of Figure 12 from (a) to (f), the most effective method for keeping track of the dynamic shift in self-care under anesthesia may be Poincare plotting. The point on the plot is represented by the value of each subsequent pair of RR intervals.

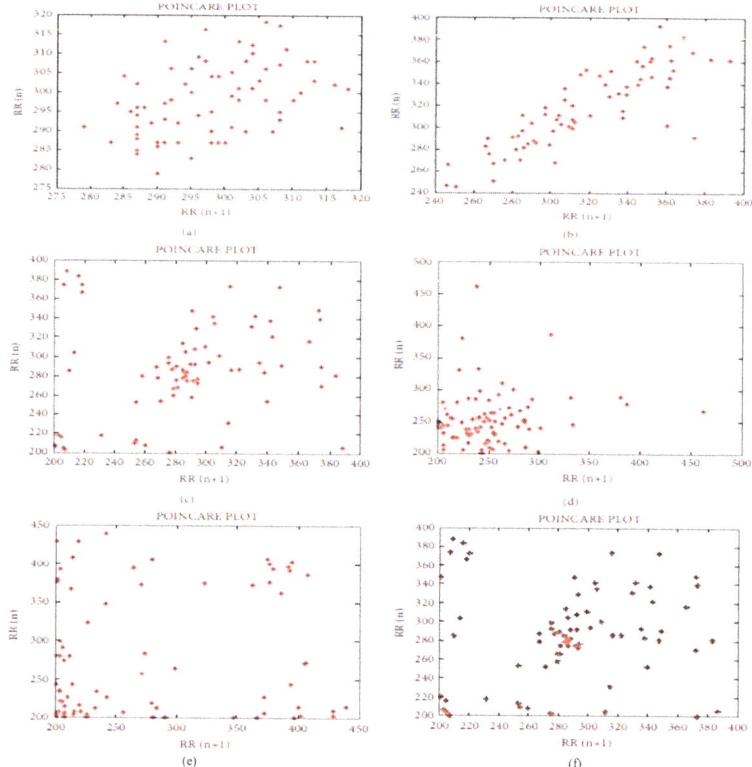

Figure 12. (a–f) Results of HRV analysis for point on the plot is represented by the value of each subsequent pair of RR intervals.

7. Conclusions

QRS peak detection helps to detect diseases and heart rate variability (HRV). The temporal history affects HRV detection. This study proposed a new ECG data classification and QRS peak detection method. The min–max method optimizes reduced-order filter coefficients. This study proposes a fuzzy-based ECG classification rule to improve performance. The projected ideal filter increases amplitude and maintains ECG characteristics, thus improving detection. The classification rule increases true positives and decreases false negatives. This study reported Poincare plot results of ECG analysis. The projected filter with the optimization method outperformed the other methods, because it achieved 100% precession and improved accuracy by 13.5% over the basic Pan–Tompkins approach and by 8.1% over the current IIR-filter-based classification rules. We hope to find more ECG signal peaks using our analysis method. The identification of ECG peaks can be used for many CVD detection applications.

Author Contributions: Conceptualization, N.M. and N.P.S.; methodology, N.M. and N.P.S.; software, N.M.; validation, N.M., N.P.S., and P.A.B.; formal analysis, N.M.; investigation, N.M.; resources, N.M.; data curation, N.M.; writing—original draft preparation, N.M.; writing—review and editing, N.M.; visualization, N.M.; supervision, N.P.S.; project administration, N.M. and N.P.S.; funding acquisition, N.P.S. and P.A.B. All authors have read and agreed to the published version of the manuscript.

Funding: This research received no external funding.

Institutional Review Board Statement: Not applicable.

Informed Consent Statement: Not applicable.

Data Availability Statement: Not applicable.

Acknowledgments: We would like to thank our supervisors for their constant encouragement and guidance toward this research work. Special thanks to Koneru Lakshmaiah Education Foundation (K L Deemed to be University) for providing institutional facilities and necessary administrative and authoritative assistance during University work.

Conflicts of Interest: The authors declare no conflict of interest.

References

1. Benmalek, M.; Charef, A. Digital fractional order operators for R-wave detection in electrocardiogram signal. *IET Signal Process.* **2009**, *3*, 381–391. [CrossRef]
2. Kaya, Y.; Pehlivan, H.; Tenekeci, M.E. Effective ECG beat classification using higher order statistic features and genetic feature selection. *J. Biomed. Res.* **2017**, *28*, 7594–7603.
3. Pandit, S.; Shukla, P.K.; Tiwari, A.; Shukla, P.K.; Maheshwari, M.; Dubey, R. Review of video compression techniques based on fractal transform function and swarm intelligence. *Int. J. Mod. Phys. B* **2020**, *34*, 2050061. [CrossRef]
4. Kaya, Y.; Pehlivan, H. Feature selection using genetic algorithms for premature ventricular contraction classification. In Proceedings of the Ninth International Conference on IEEE Electrical and Electronics Engineering, Bursa, Turkey, 26–28 November 2015; pp. 1229–1232.
5. Shukla, P.K.; Sandhu, J.K.; Ahirwar, A.; Ghai, D.; Maheshwary, P.; Shukla, P.K. Multiobjective Genetic Algorithm and Convolutional Neural Network Based COVID-19 Identification in Chest X-ray Images. *Math. Probl. Eng.* **2021**, *2021*, 7804540. [CrossRef]
6. Manikandan, M.S.; Seaman, K.P. A novel method for detecting R-peaks in electrocardiogram (ECG) signal. *J. Biomed. Signal Process. Control.* **2012**, *7*, 118–128. [CrossRef]
7. Roy, V.; Shukla, S. Designing Efficient Blind Source Separation Methods for EEG Motion Artifact Removal Based on Statistical Evaluation. *Wirel. Pers. Commun.* **2019**, *108*, 1311–1327. [CrossRef]
8. Kaur, H.; Rajni, R. On the detection of Cardiac Arrhythmia with Principal Component Analysis. *Wirel. Pers. Commun.* **2017**, *97*, 5495–5509. [CrossRef]
9. Padmavathi, K.; Ramakrishna, K.S. Classification of ECG Signal during Atrial Fibrillation Using Autoregressive Modeling. *Procedia Comput. Sci.* **2015**, *46*, 53–59. [CrossRef]
10. Roy, V.; Shukla, P.K.; Gupta, A.K.; Goel, V.; Shukla, P.K.; Shukla, S. Taxonomy on EEG Artifacts removal methods, issues, and healthcare applications. *J. Organ. End User Comput.* **2021**, *33*, 19–46. [CrossRef]
11. Peterkova, A.; Stremy, M. The raw ECG signal processing and the detection of QRS complex. In Proceedings of the IEEE European Modelling Symposium, Madrid, Spain, 6–8 October 2015.
12. Pan, J.; Tompkins, W.J. A real time QRS detection algorithm. *IEEE Trans. Biomed. Eng.* **1985**, *32*, 230–236. [CrossRef]
13. Roy, V.; Shukla, S.; Shukla, P.K.; Rawat, P. Gaussian Elimination-Based Novel Canonical Correlation Analysis Method for EEG Motion Artifact Removal. *J. Health Eng.* **2017**, *2017*, 9674712. [CrossRef] [PubMed]
14. Verma, V.; Rathore, S.S. Comparative study of QRS complex detection by threshold technique. *Int. J. Adv. Eng. Technol.* **2015**, *8*, 22311963.
15. Salih, S.K.; Aljunid, S.A.; Yahya, A.; Ghailan, K.Y. A novel approach for detecting QRS complex of ECG signal. *Int. J. Comput. Sci. Issues* **2012**, *9*, 205.
16. Roy, V.; Shukla, S. A methodical health-care model to eliminate motion artifacts from big EEG data, JOEUC, big data analytics in business. *Healthc. Gov.* **2016**, *29*, 1546–2234.
17. Sharma, T.; Sharma, K.K. QRS complex detection in ECG signals using the synchros queezed wavelet transform. *IETE J. Res.* **2016**, *62*, 885–892. [CrossRef]
18. Naaz, A.; Singh, M. QRS complex detection and ST segmentation of ECG signal using wavelet transform. *Int. J. Res. Advent Technol.* **2015**, *3*, 45–50.
19. Moody, G.B.; Mark, R.G. The impact of the MIT-BIH Arrhythmia Database. *IEEE Eng. Med. Biol. Mag.* **2001**, *20*, 45–50. [CrossRef]
20. Dohare, A.K.; Kumar, V.; Kumar, R. An efficient new method for the detection of QRS in electrocardiogram. *Comput. Electr. Eng.* **2014**, *40*, 1717–1730. [CrossRef]
21. Kaur, I.; Rajni, R.; Marwaha, A. ECG Signal Analysis and Arrhythmia Detection using Wavelet Transform. *J. Inst. Eng. Ser. B* **2016**, *97*, 499–507. [CrossRef]
22. Benitez, D.; Gaydecki, P.A.; Zaidi, A.; Fitzpatrick, A.P. +e use of the Hilbert transform in ECG signal analysis. *Comput. Biol. Med.* **2001**, *31*, 399–406. [CrossRef]
23. Hamilton, P.S.; Tompkins, W.J. Quantitative investigation of QRS detection rules using the MIT/BIH arrhythmia database. *IEEE Trans. Biomed. Eng.* **1986**, *33*, 1157–1165. [CrossRef] [PubMed]
24. Sun, Y.; Chan, K.L.; Krishnan, S.M. Characteristic wave detection in ECG signal using morphological transform. *BMC Cardiovasc. Disord.* **2005**, *5*, 28. [CrossRef] [PubMed]

25. Sargar, L.S.; Gharat, M.M.; Bhat, S.N.; Bagal, U.R. Automated detection of R-peaks in electrocardiogram. *Int. J. Sci. Eng. Res.* **2015**, *6*, 1265–1269.
26. Qin, Q.; Li, J.; Yue, Y.; Liu, C. An adaptive and time efficient ECG R-peak detection algorithm. *J. Healthcare Eng.* **2017**, *2017*, 14. [CrossRef] [PubMed]
27. Slimane, Z.H.; Ali, A.N. QRS complex detection using empirical mode decomposition. *Digit. Signal Process.* **2010**, *20*, 1221–1228. [CrossRef]
28. Sivakumar, R.; Tamilselvi, R.; Abinaya, S. Noise analysis & QRS detection in ECG signals. *Int. Conf. Comput. Technol. Sci.* **2012**, *47*, 141–146.
29. Khan, M.T.; Ahamed, S.R. A New High Performance VLSI Architecture for LMS Adaptive Filter Using Distributed Arithmetic. In Proceedings of the 2017 IEEE Computer Society Annual Symposium on VLSI (ISVLSI), Bochum, Germany, 3–5 July 2017; IEEE: New York, NY, USA; pp. 219–224.
30. Haykin, S.; Widrow, B. *Least-Mean-Square Adaptive Filters*; Wiley-Interscience: Hoboken, NJ, USA, 2003.
31. Allred, D.; Yoo, H.; Krishnan, V.; Huang, W.; Anderson, D. LMS adaptive filters using distributed arithmetic for high throughput. *IEEE Trans. Circuits Syst. I Regul. Pap.* **2005**, *52*, 1327–1337. [CrossRef]
32. Krad, H.; Al-Taie, A.Y. Performance Analysis of a 32-Bit Multiplier with a Carry-Look-Ahead Adder and a 32-bit Multiplier with a Ripple Adder using VHDL. *J. Comput. Sci.* **2008**, *4*, 305–308. [CrossRef]
33. Mottaghi-Dastjerdi, M.; Afzali-Kusha, A.; Pedram, M. BZ-FAD: A Low-Power Low-Area Multiplier Based on Shift-and-Add Architecture. *IEEE Trans. Very Large Scale Integr. (VLSI) Syst.* **2009**, *17*, 302–306. [CrossRef]
34. Saha, A.; Pal, D.; Chandra, M. Low-power 6-GHz wave-pipelined 8b × 8b multiplier. *IET Circuits Devices Syst.* **2013**, *7*, 124–140. [CrossRef]

MDPI

Proceeding Paper

Implementation of Content-Based Image Retrieval Using Artificial Neural Networks [†]

Sarath Chandra Yenigalla *, Karumuri Srinivasa Rao and Phalguni Singh Ngangbam

Multi-Core Architecture Computation (MAC) Lab, Department of Electronics and Communication Engineering, K L University (Deemed to be University), Vijayawada 522501, India

* Correspondence: sarathcy2000@gmail.com

† Presented at the International Conference on "Holography meets Advanced Manufacturing", Online, 20–22 February 2023.

Abstract: CBIR (Content Based Image Retrieval) has become a critical domain in the previous decade, owing to the rising demand for image retrieval from multimedia databases. Typically, we take low-level (colour, texture and shape) or high-level (when machine learning techniques are used) features out of the photos. In our research, we examine the CBIR system utilising three machine learning methods, namely SVM (Support Vector Machine), KNN (K Nearest Neighbours), and CNN (Convolution Neural Networks), using Corel 1K, 5K, and 10K databases, by splitting the data into 80% train data and 20% test data. Moreover, compare each algorithm's accuracy and efficiency when a specific task of image retrieval is given to it. The final outcome of this project will provide us with a clear vision of how effective deep learning, KNN and CNN algorithms are to finish the task of image retrieval.

Keywords: Content Based Image Retrieval; Convolution Neural Networks; deep learning

1. Introduction

1.1. Image Retrieval

The explosive growth of digital images in recent years has led to the development of image retrieval systems The viewing, searching, and retrieval of images from a sizable database of digital photographs are made possible by image retrieval systems. Adding metadata to images, such as captions, keywords, titles, or descriptions, is a common practise in traditional techniques of image retrieval. To address this challenge, extensive research has been conducted on automatic image annotation.

The design of web-based picture annotation tools has been influenced by both conventional approaches as well as the growth of social web apps and the semantic web. Image retrieval search techniques include content-based image retrieval (CBIR), image collection exploration, and image meta-search. A user-supplied query picture or user-specified image features are used by CBIR instead of written descriptions to determine how similar an image's contents, such as textures, colours, and forms, are to the query image.

It is critical to establish the extent and nature of picture data in order to assess the degree of complexity of the image search system architecture. A search system's predicted user traffic and the user base's diversity are two more elements that affect design.

1.2. Content-Based Image Retrieval

The task of finding digital images in massive databases is known as the "image retrieval problem", and content-based image retrieval (CBIR) is the practical application of machine learning algorithms to this problem. Query by image content (QBIC) is another name for CBIR [1]. Traditional concept-based methods that rely on metadata such as keywords, tags, or image descriptions are not supported by CBIR. The term "content" in

Citation: Yenigalla, S.C.; Rao, K.S.; Ngangbam, P.S. Implementation of Content-Based Image Retrieval Using Artificial Neural Networks. *Eng. Proc.* **2023**, *34*, 25. https://doi.org/10.3390/HMAM2-14161

Academic Editor: Vijayakumar Anand

Published: 13 March 2023

this case can refer to colours, shapes, textures, or any additional data that can be inferred from the image itself. The search studies the contents of the image rather than its metadata.

CBIR is advantageous since searches that just employ metadata are reliant on the accuracy and completeness of the annotations, and because manually annotating photos by inserting keywords or other metadata in a big database can be difficult to execute and may not yield the required results. Similar to keyword image search, which is arbitrary and poorly defined, CBIR systems face difficulties in quantifying success. IBM built the first commercial CBIR system, QBIC (Query by Image Content), and newer network and graph-based systems have presented easy and appealing replacements to existing methods.

Due to the limits of metadata-based systems and the wide variety of applications for effective image retrieval, CBIR has increased interest. To address these needs, user-friendly interfaces and human-centred design have begun to be included in CBIR research. Many other features are now employed in CBIR systems, which were first designed to search databases using picture properties including colour, texture, and shape. However, scalability and miscategorisation problems persist with standards created to classify photos.

CBIR has been applied to a variety of applications, including satellite imagery [2], mapping, medical imaging [3], fingerprint scanning [4,5], and biodiversity information systems. The overall goal of this research article is to investigate content-based picture retrieval including its approaches, strategies, and uses. The paper also covers current research initiatives, difficulties, and CBIR's future directions.

2. Methodology

The proposed CBIR (Content-Based Image Retrieval) system with machine learning consists of an offline and online phase. In the offline phase, the system extracts feature vectors using Local Patterns methods for all images in the database, labels 60–70% of images from each class [6], and trains a machine learning classifier (e.g., SVM, KNN, and CNN) [7] to predict class names for each feature vector. In the online phase, the user inputs a query image, its feature vector is calculated using LNP (Local Neighbour Pattern), and the machine learning classifier predicts the class name. The system retrieves images from the same class in the offline phase using Euclidean distance calculations and presents the top K results to the user. Three datasets were used to test the system: Corel 1K (1000 images with 10 classes of 100 images each), Vistex (640 images of size 512×512), and Faces (40 classes, each with 10 images of size 112×92 pixels, showing variations in lighting, facial details, and expressions). Figure 1 shows the architecture of Content-Based Image Retrieval.

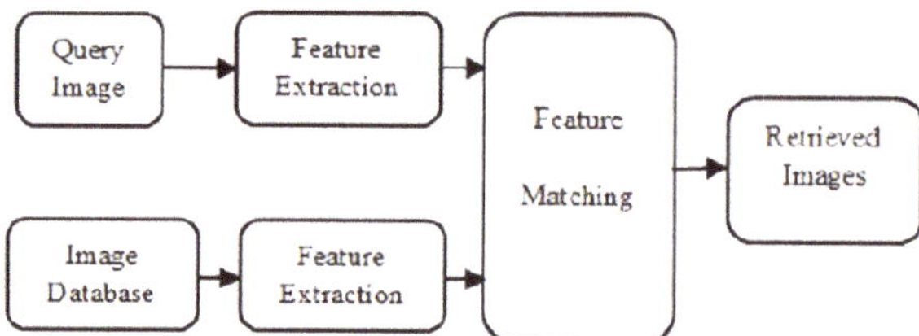

Figure 1. CBIR architecture.

2.1. Co-Occurrence Matrix Calculation

Suppose the input image has Nc and Nr pixels in the horizontal and vertical directions, respectively. Assume $Zc = \{1, 2, \dots, Nc\}$ is a horizontal space domain and $Zr = \{1, 2, \dots, Nr\}$ is a vertical space domain. When the direction θ and distance d are given, the matrix element $P(i,j/d,\theta)$ can be expressed by calculating the pixel logarithm of co-occurrence grey levels i and j. Assume the distance is 1, θ equals $0°$, $45°$, $90°$, and $135°$, respectively, the formulae are:

$$P(i,j/1,0) = \#\{[(k,l),(m,n)] \in (Zr \times Zc) \mid k - m \mid = 0, \mid l - n \mid = 1, f(k,l) = i, f(m,n) = j\}, \quad (1)$$

$$P(i,j/1,90) = \#\{[(k,l),(m,n)] \in (Zr \times Zc) \mid k - m \mid = 1, \mid l - n \mid = 0, f(k,l) = i, f(m,n) = j\}, \quad (2)$$

$$P(i,j/1,45) = \#\{[(k,l),(m,n)] \in (Zr \times Zc)(k - m) = 1,(l - n) = -1 \text{ or}(k - m) = -1,(l - n) = 1, f(k,l) = i, f(m,n) = j\}, \quad (3)$$

$$P(i,j/1,135) = \#\{[(k,l),(m,n)] \in (Zr \times Zc)(k - m) = 1,(l - n) = 1 \text{ or}(k - m) = -1,(l - n) = -1, f(k,l) = i, f(m,n) = j\}, \quad (4)$$

where # represents the pixel logarithm, which generates brackets, and k,m and l,n reflect modifications of selected calculation windows.

2.2. Texture Features Extraction

Formula (5) will convert a colour image into a 256-level greyscale image.

$$Y = 0.114 \times B + 0.587 \times G + 0.29 \times R, \quad (5)$$

where Y is the grey-scale value. R, G, and B represent red, green, and blue component values, respectively. Because the grey scale is 256, the corresponding co-occurrence matrix is 256×256. The grey scale of the initial image will be compressed to reduce calculations before the co-occurrence matrix is formed. A total of 16 compression levels were chosen in the paper to improve the texture feature extracting speed. Four co-occurrence matrices are formed according to Formula (3) to Formula (6) in four directions. The four texture parameters are calculated: capacity, entropy, moment of inertia, and relevance.

For an image li and its corresponding feature vector Hi = [hi, 1, hi, 2, ... , hi, N], assume the feature component value satisfies a Gaussian distribution. The Gaussian normalization approach is used to implement internal normalization in order to make each feature of the same weight.

$$h_{i,j}' = h_{i,j} - m_j\sigma_j, \quad (6)$$

where mj is mean and σj is the standard deviation. hi,j will be unitized on a range $[-1,1]$. The texture feature of each image is calculated according to the above steps. The texture values are compared by Euclidean distance, the closer the distance the higher the similarity.

3. Results and Discussion

Using average precision, the proposed retrieval system's effectiveness is assessed for each query. Equation (7) can be used to obtain the area beneath each query's precision, where the precision is the proportion of relevant images to all images retrieved.

$$\text{Precision} = \frac{relevant\ images - retrieved\ images}{retrieved\ images} \quad (7)$$

In this Proposed method, three different types of databases area used, and three types of techniques are used. Here, the query images (k) are fixed at 8. Figures 2–4 show the retrieved images in an animal database and the accuracy of the image with respect to the searched image is shown below the respective image.

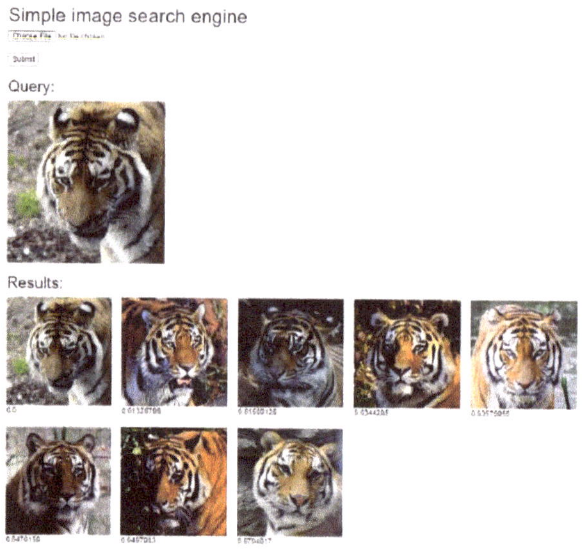

Simple image search engine

Query:

Results:

Figure 2. Retrieved animal images.

Simple image search engine

Query:

Results:

Figure 3. Retrieved bus images.

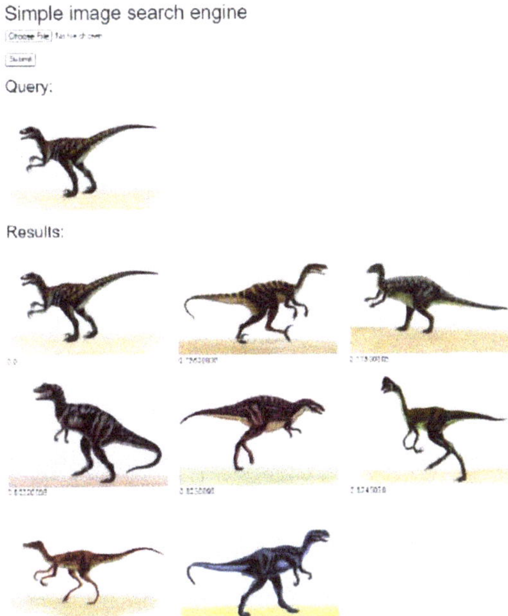

Figure 4. Retrieved dinosaur images.

Figures 5–7 show the average accuracy results of pixel and histogram accuracy both with respect to the algorithm used.

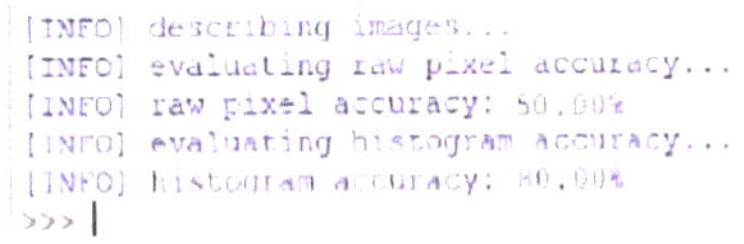

```
[INFO] describing images...
[INFO] evaluating raw pixel accuracy...
[INFO] raw pixel accuracy: 50.00%
[INFO] evaluating histogram accuracy...
[INFO] histogram accuracy: 80.00%
>>> |
```

Figure 5. CNN Accuracy Output.

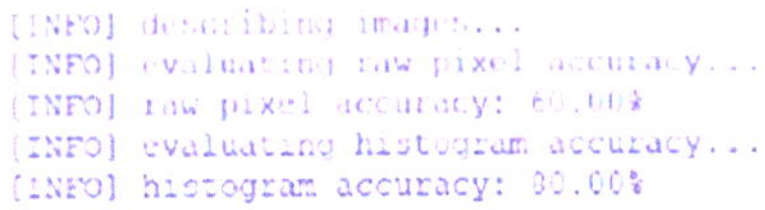

```
[INFO] describing images...
[INFO] evaluating raw pixel accuracy...
[INFO] raw pixel accuracy: 60.00%
[INFO] evaluating histogram accuracy...
[INFO] histogram accuracy: 90.00%
```

Figure 6. KNN Accuracy Output.

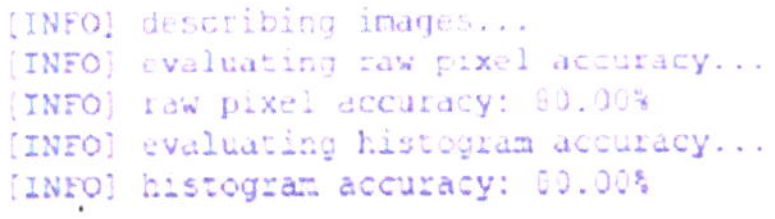

```
[INFO] describing images...
[INFO] evaluating raw pixel accuracy...
[INFO] raw pixel accuracy: 80.00%
[INFO] evaluating histogram accuracy...
[INFO] histogram accuracy: 80.00%
```

Figure 7. SVM Accuracy Output.

Table 1 shows the experimental results of images retrieved depending upon the accuracy of the algorithm in which the CNN technique has the least 50% and KNN has 60% and SVM has 80%.

Table 1. Accuracy of Algorithms using 3 Databases.

Algorithm	Corel 1K	Animal	Dinosaur
KNN	60%	60%	60%
CNN	50%	50%	50%
SVM	80%	80%	80%

4. Conclusions

Following a review of prior CBIR efforts, the paper investigated the low-level aspects of CBIR colour and texture extraction. The paper created a CBIR system using fused characteristics of colour and texture after testing the three distinct types of algorithms (KNN, CNN, and SVM) with various databases (Corel 1K, Animal, and Dinosaur). By entering a query image, similar photographs can be correctly and quickly retrieved. Moreover, the above works discuss the calculation of the accuracy of each algorithm with all the three databases and conclude that the CNN algorithm has 50% accuracy with all three databases, the KNN algorithm has 60% accuracy with all three databases, and the SVM algorithm has 80% accuracy with all the three databases, and so finally we came to a conclusion that SVM is the best suit algorithm among the three algorithms with an accuracy of 80%.

In the future, more low-level features will be combined to strengthen the system, such as spatial position and shape features. The other two key components of the CBIR system are the image feature matching approach and semantic-based image retrieval.

Author Contributions: S.C.Y.: conceptualization, methodology, software, investigation, writing—original draft preparation, writing—review and editing, formal analysis. S.R.: project administration, resources, investigation, and methodology. P.S.N.: supervision. All authors have read and agreed to the published version of the manuscript.

Funding: This research received no external funding.

Institutional Review Board Statement: Not applicable.

Informed Consent Statement: Not applicable.

Data Availability Statement: No data was used for the research described in the article.

Conflicts of Interest: The authors declare no conflict of interest.

References

1. Sharma, S. Use of Artificial Intelligence Algorithm for Content-Based Image Retrieval Syste. *Int. J. Adv. Res. Ideas Innov. Technol.* **2018**, *4*, 680–684.
2. Ferran, A.; Bernabe, S.; Rodriguez, P.; Plaza, A. A Web-Based System for Classification of Remote Sensing Data. *IEEE J. Sel. Top. Appl. Earth Obs. Remote Sens.* **2013**, *6*, 1934–1948. [CrossRef]
3. Ramos, J.; Kockelkorn, T.; Ramos, I.; Ramos, R.; Grutters, J.; Viergever, M.; van Ginneken, B.; Campilho, A. Content-Based Image Retrieval by Metric Learning From Radiology Reports: Application to Interstitial Lung Diseases. *IEEE J. Biomed. Health Inform.* **2016**, *20*, 281–292. [CrossRef] [PubMed]
4. Zegarra, J.M.; Leite, N.; da Silva Torres, R. Wavelet-based fingerprint image retrieval. *J. Comput. Appl. Math.* **2009**, *227*, 294–307. [CrossRef]
5. Gavrielides, M.; Sikudova, E.; Pitas, I. Color-based descriptors for image fingerprinting. *IEEE Trans. Multimed.* **2006**, *8*, 740–748. [CrossRef]
6. Alrahhal, M.; Supreethi, K.P. Content-Based Image Retrieval using Local Patterns and Supervised Machine Learning Techniques. In Proceedings of the 2019 Amity International Conference on Artificial Intelligence (AICAI), Dubai, United Arab Emirates, 4–6 February 2019; pp. 118–124. [CrossRef]
7. Wiggers, K.L.; Britto, A.S.; Heutte, L.; Koerich, A.L.; Oliveira, L.E.S. Document Image Retrieval Using Deep Features. In Proceedings of the 2018 International Joint Conference on Neural Networks (IJCNN), Rio de Janeiro, Brazil, 8–13 July 2018; pp. 1–8. [CrossRef]

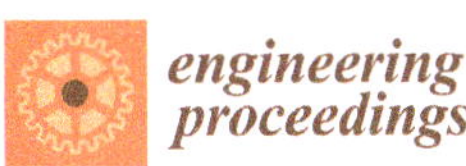

Proceeding Paper

Imaging with Diffractive Axicons Rapidly Milled on Sapphire by Femtosecond Laser Ablation [†]

Daniel Smith [1,*], Soon Hock Ng [1], Molong Han [1], Tomas Katkus [1], Vijayakumar Anand [2,*] and Saulius Juodkazis [1,3]

[1] Optical Sciences Centre and ARC Training Centre in Surface Engineering for Advanced Materials (SEAM), School of Science, Computing and Engineering Technologies, Swinburne University of Technology, Hawthorn, VIC 3122, Australia; soonhockng@swin.edu.au (S.H.N.); molonghan@swin.edu.au (M.H.); tkatkus@swin.edu.au (T.K.); sjuodkazis@swin.edu.au (S.J.)

[2] Institute of Physics, University of Tartu, Ülikooli 18, 50090 Tartu, Estonia

[3] World Research Hub Initiative (WRHI), School of Materials and Chemical Technology, Tokyo Institute of Technology, 2-12-1, Ookayama, Meguro-ku, Tokyo 152-8550, Japan

[*] Correspondence: danielsmith@swin.edu.au (D.S.); vanand@swin.edu.au (V.A.)

[†] Presented at the International Conference on "Holography Meets Advanced Manufacturing", Online, 20–22 February 2023.

Abstract: We show that single-pulse burst fabrication will produce a flatter and smoother profile of axicons milled on sapphire compared to pulse overlapped fabrication which results in a damaged and much rougher surface. The fabrication of large-area (sub-1 cm cross-section) micro-optical components in a short period of time ($\sim$10 min) and with less processing steps is highly desirable and would be cost-effective. Our results were achieved with femtosecond laser fabrication technology which has revolutionized the field of advanced manufacturing. This study compares three configurations of axicons such as the conventional axicon, a photon sieve axicon (PSA) and a sparse PSA directly milled onto a sapphire substrate. Debris of redeposited amorphous sapphire were removed using isopropyl alcohol and potassium hydroxide. A spatially incoherent illumination was used to test the components for imaging applications. Non-linear reconstruction was used for cleaning noisy images generated by the axicons.

Keywords: ablation; femtosecond lasers; diffractive optical elements; imaging; astronomy

Citation: Smith, D.; Ng, S.H.; Han, M.; Katkus, T.; Anand, V.; Juodkazis, S. Imaging with Diffractive Axicons Rapidly Milled on Sapphire by Femtosecond Laser Ablation. *Eng. Proc.* **2023**, *34*, 26. https://doi.org/10.3390/HMAM2-14147

Academic Editor: Kaupo Kukli

Published: 13 March 2023

1. Introduction

Axicons are diffractive optical elements which have a line focus. Importantly. axicons are Bessel beam generators and can be used for a range of different applications including optical trapping [1,2]. Bessel beams are important as they are diffraction resistant and can be used for optical imaging applications [3] and biomedical applications such as ultrasound imaging [4–6]. (Figure 1) shows the lens axicon with its unique large depth of focus (DOF) and Bessel beam generation.

In this body of work, an optimal and record processing time for micro-optical components is described, where the process used has fewer steps and is more cost effective than previous attempts. Instead of using axicon lenses, micro-optical binary surface axicons with a two phase or amplitude configurations are used. The extra configurations give rise to, in total, three different diffractive optical elements: a conventional surface axicon, surface photon sieve axicon (PSA), and sparse surface photon sieve axicon (sparse PSA) which were directly milled onto a sapphire substrate.

This particular micro-optical component falls into the category of diffractive optics and plays an important role in many areas of research; furthermore, they have been manufactured using different techniques. The techniques generally employed are photolithography [7], electron beam lithography [8], and ion beam lithography [9], which are time-consuming and expensive, making the device itself expensive. Femtosecond laser ablation is superior to all of the above methods with only one exception: the ability to deliver a practical solution for large areas such as millimeter-to-centimeter-scale micro-optical element fabrication [10–13]. Lately, there has been a shift in focus in imaging research from using coherent light sources to incoherent ones due to the many advantages of the latter, such as their broad applicability, low cost, and high resolving power. We have used femtosecond fabrication with ablation to rapidly fabricate two-level axicons directly onto sapphire substrates. Two configurations, conventional ring and sieve configurations, were produced. Furthermore, our work focused on rapid fabrication to manufacture large-area diffractive optical elements for astronomical imaging applications [14].

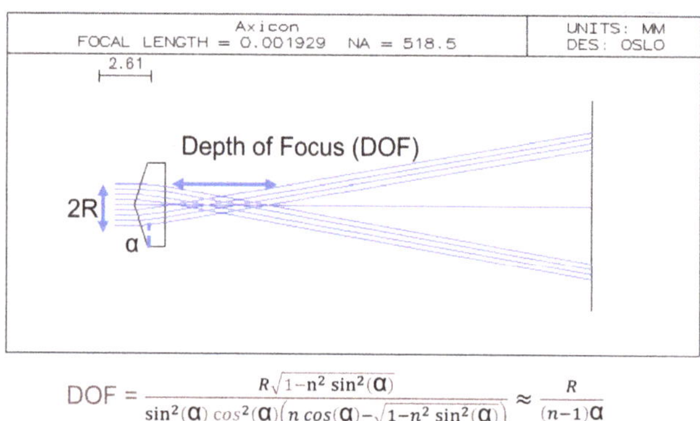

$$DOF = \frac{R\sqrt{1-n^2\sin^2(\alpha)}}{\sin^2(\alpha)\cos^2(\alpha)\left(n\cos(\alpha)-\sqrt{1-n^2\sin^2(\alpha)}\right)} \approx \frac{R}{(n-1)\alpha}$$

Figure 1. Conventional axicon lens simulated in OSLO software for application at a wavelength of 550 nm. Depth of focus equation for conventional axicon lens.

The beam characteristics were investigated for a spatially incoherent illumination since it is available at low cost and is therefore relevant to large-scale astronomical applications. In astronomy, large area devices are required which can be easily manufactured using femtosecond fabrication systems. The optical configuration of light diffracted from a point object is incident on a diffractive axicon and the intensity distribution is recorded (shown in Figure 2).

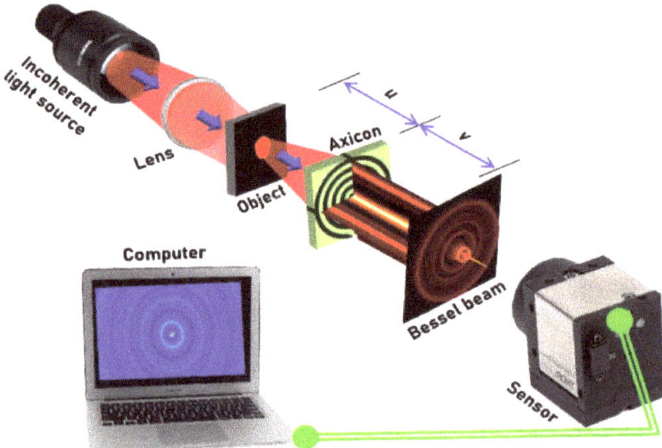

Figure 2. Optical configuration for the generation of Bessel beams by a diffractive axicon. u is the distance from the object being imaged, and v is the distance between the imaging axicon to the sensor. In this case u = v. Reprinted with permissions from Ref. [14]. Copyright 2022 Applied Physics B.

2. Results and Discussion: Laser Ablation and Imaging

2.1. Laser Ablation

The fabrication was carried out on a sapphire substrate of 500 μmin thickness and 10 mm by 10 mm in size. Sapphire has a refractive index of $n_{sapphire} = 1.76$ and has a spectral window from 300 nm to 6000 nm. The milling depth, t, is determined by the relationship between the wavelength that the diffractive optical element is desired to interact with and the refractive index of the material ($t = \lambda/2(n_{sapphire} - 1)$). The thickness t was determined to be 400 μm for $\lambda \approx 600$ nm. The wavelength source for optical testing is 617 nm. Acetone and Iso-Propyl alcohol were used to clean the substrate prior to fabrication.

Three optical devices—a conventional surface axicon, surface photon sieve axicon (PSA), and sparse surface photon sieve axicon (sparse PSA), shown in Figure 3a,b,c respectively—were created using the light conversion PHAROS laser operating at a 200 kHz repetition rate, λ = 1030 nm wavelength, 2.5 W average power, 230 fs pulse duration and a 5× magnification, NA = 0.14 numerical aperture Mitutoyo Plan APO NIR infinity corrected objective lens. Two pulses per ablation spot were used to ablate the regions of the axicon; as a consequence, these more heavily overlapped regions incurred ripple formation. The attenuator serves to create more precise beam energies in combination with the objective lens. Here, a 5 μJ beam was used to achieve the fabrication which led to pulse energies of 70 TW/cm^2, even though the ablation threshold of sapphire is only in the order of 10 TW/cm^2 and with ablation spots with a diameter of 8.9 μm.

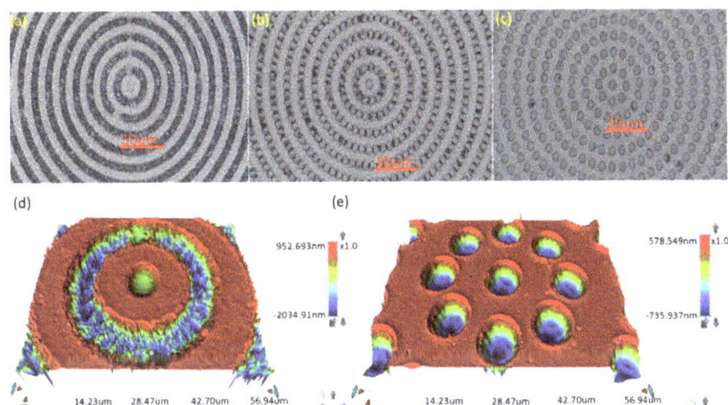

Figure 3. Optical microscope images of the three different devices created: a conventional surface axicon (**a**), surface photon sieve axicon (PSA) (**b**), and sparse surface photon sieve axicon (sparse PSA) (**c**). Below we provide the optical profilometer images of the conventional surface axicon (**d**) and sparse surface photon sieve axicon (sparse PSA) (**e**). Reprinted with permissions from Ref. [14]. Copyright 2022 Applied Physics B.

2.2. Imaging

Optical testing was carried out according to Figure 2 using a high-power LED and a spectral filter to improve the temporal coherence. A pinhole with a size of 100 μm and a cross shaped object Figure 4a,b were used for imaging. A 3× magnifying system was used to re-image the intensity distribution that is close to the diffractive optical elements on an image sensor. These optical elements were mounted one after the other and the intensity distributions were recorded at 5 mm for the axicon, PSA and sparse PSA. The pinhole was replaced by a cross object and the intensity patterns again were recorded at 3 cm from the DOEs. Again, the images were cleaned using non-linear reconstruction and to further improve the cleaning results, additional filters such as a median filter and correlation filters were used. A significant difference was seen in the cleaned images compared to the original images without filters.

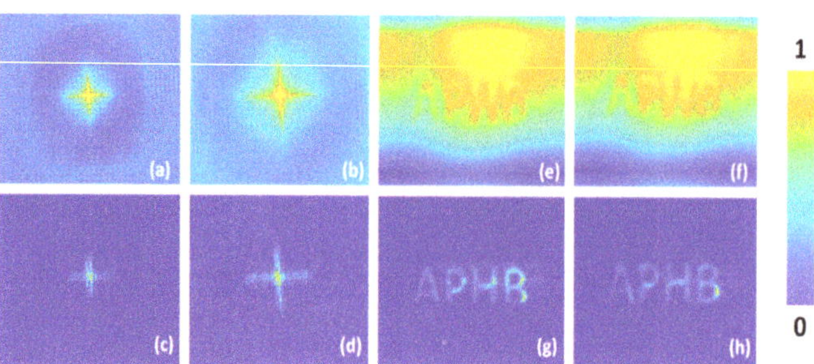

Figure 4. Intensity distribution of cross object recorded at (**a**) z = 5 cm and (**b**) z = 6 cm and the corresponding cleaned images (**c**) z = 5 cm and (**d**) z = 6 cm, respectively. Intensity distribution of synthetic object 'APHB' recorded at (**e**) z = 5 cm and sparse (**f**) z = 6 cm and the corresponding cleaned images (**g**) z = 5 cm and (**h**) z = 6 cm, respectively. Reprinted with permissions from Ref. [14]. Copyright 2022 Applied Physics B.

3. Conclusions and Future Works

This study showed that the fabrication time could be cut to 10 min using the femtosecond laser fabrication method for large-area, i.e., 5 mm × 5 mm, objects. It was noticed that the beam overlap during milling resulted in the redeposition of material resulting in a lower depth than the case without beam overlap, which is contrary to the common belief that overall higher exposure with the beam overlap increases the depth. An increase in roughness has been noticed and it is contributed to by the redeposition and light-matter interaction at temperature changes due to the ablation process. It is noted that debris will have to be reduced to avoid a build-up of rough areas. Some of the latest developments in astronomical spectral imaging technologies such as FOBOS require numerous micro-optical devices for the successful implementation of free space to fiber bundle coupling for spectral imaging. The current work shows the possibility of rapid fabrication and beam cleaning capable of retrieving spatial information in addition to the recorded spectral information.

Author Contributions: Conceptualisation, S.J., V.A. and S.H.N.; investigation, D.S., M.H., V.A. and T.K.; writing— original draft preparation, D.S.; review and editing, all authors. All authors have read and agreed to the published version of the manuscript.

Funding: This research was funded by the Australian Research Council, grant number LP190100505. S.J. is grateful for startup funding from the Nanotechnology Facility at Swinburne and to the Workshop-on-Photonics for the technology transfer project, which installed the first industrial grade microfabrication setup in Australia. Part of this study was created during a Grand Challenge undergraduate project of A.T.

Institutional Review Board Statement: Not applicable.

Informed Consent Statement: Not applicable.

Data Availability Statement: The data presented in this study are available on request from the corresponding author.

Conflicts of Interest: The authors declare no conflict of interest.

References

1. Ding, Z.; Lai, G. Enhancement of axial optical trapping force using a pair of axicons. In Proceedings of the Technical Digest. CLEO/Pacific Rim '99. Pacific Rim Conference on Lasers and Electro-Optics (Cat. No.99TH8464), Seoul, Republic of Korea, 30 August–3 September 1999; Volume 2, pp. 369–370. [CrossRef]
2. Balasubramani, V.; Vijayakumar, A.; Rai, M.R.; Rosen, J.; Cheng, C.J.; Minin, O.V.; Minin, I.V. Binary square axicon with chiral focusing properties for optical trapping. *Opt. Eng.* **2019**, *59*, 041204. [CrossRef]
3. Khonina, S.N.; Kazanskiy, N.L.; Karpeev, S.V.; Butt, M.A. Bessel Beam: Significance and Applications—A Progressive Review. *Micromachines* **2020**, *11*, 997. [CrossRef] [PubMed]

4. Hsu, D.K.; Margetan, F.J.; Thompson, D.O. Bessel beam ultrasonic transducer: Fabrication method and experimental results. *Appl. Phys. Lett.* **1989**, *55*, 2066–2068. [CrossRef]
5. Lu, J.; Song, T.; Kinnick, R.; Greenleaf, J. In vitro and in vivo real-time imaging with ultrasonic limited diffraction beams. *IEEE Trans. Med. Imaging* **1993**, *12*, 819–829. [CrossRef] [PubMed]
6. Lu, J.Y.; Xu, X.L.; Zou, H.; Greenleaf, J. Application of Bessel beam for Doppler velocity estimation. *IEEE Trans. Ultrason. Ferroelectr. Freq. Control* **1995**, *42*, 649–662. [CrossRef]
7. Kondo, T.; Juodkazis, S.; Mizeikis, V.; Misawa, H.; Matsuo, S. Holographic lithography of periodic two-and three-dimensional microstructures in photoresist SU-8. *Opt. Express* **2006**, *14*, 7943–7953. [CrossRef]
8. Anand, V.; Ng, S.H.; Katkus, T.; Juodkazis, S. White light three-dimensional imaging using a quasi-random lens. *Opt. Express* **2021**, *29*, 15551–15563. [CrossRef] [PubMed]
9. Seniutinas, G.; Gervinskas, G.; Anguita, J.; Hakobyan, D.; Brasselet, E.; Juodkazis, S. Nanoproximity direct ion beam writing. *Nanofabrication* **2016**, *2*, 6. [CrossRef]
10. Marcinkevičius, A.; Juodkazis, S.; Watanabe, M.; Miwa, M.; Matsuo, S.; Misawa, H.; Nishii, J. Femtosecond laser-assisted three-dimensional microfabrication in silica. *Opt. Lett.* **2001**, *26*, 277–279. [CrossRef] [PubMed]
11. Malinauskas, M.; Žukauskas, A.; Hasegawa, S.; Hayasaki, Y.; Mizeikis, V.; Buividas, R.; Juodkazis, S. Ultrafast laser processing of materials: From science to industry. *Light. Sci. Appl.* **2016**, *5*, 2047–7538. [CrossRef] [PubMed]
12. Juodkazis, S.; Yamasaki, K.; Mizeikis, V.; Matsuo, S.; Misawa, H. Formation of embedded patterns in glasses using femtosecond irradiation. *Appl. Phys. A* **2004**, *4*, 1549–1553. [CrossRef]
13. Vanagas, E.; Kudryashov, I.; Tuzhilin, D.; Juodkazis, S.; Matsuo, S.; Misawa, H. Surface nanostructuring of borosilicate glass by femtosecond nJ energy pulses. *Appl. Phys. Lett.* **2003**, *82*, 2901–2903. [CrossRef]
14. Smith, D.; Ng, S.H.; Han, M.; Katkus, T.; Anand, V.; Glazebrook, K.; Juodkazis, S. Imaging with diffractive axicons rapidly milled on sapphire by femtosecond laser ablation. *Appl. Phys. B* **2022**, *127*, 154. [CrossRef]

engineering proceedings

Proceeding Paper

Light for Life—Optical Spectroscopy in Clinical Settings [†]

Shree Krishnamoorthy

Biophotonics@Tyndall, Irish Photonics Integration Center, Tyndall National Institute, T12 R5CP Cork, Ireland;
shree.krishnamoorthy@tyndall.ie

† Presented at the International Conference on "Holography Meets Advanced Manufacturing", Online, 20–22 February 2023.

Abstract: Diagnosing diseases in our bodies requires the measurement of physiological biomarkers non-invasively. Assessing biomarker levels is a key step in this process. Light allows for non-invasive assessment of disease in tissue. Here, I give my perspective on the use of light for diagnosis with examples of research conducted by our research team, focusing on two conditions: oral cancer and fetal hypoxia diagnosis. In the case of oral cancer, we look at the spatially localized diagnosis of cancer tissue in the oral cavity. In the case of fetal hypoxia, we look at temporal changes in physiological conditions for diagnosis. In both cases, we see the potentially transformative impact of optical spectroscopy on clinical diagnosis.

Keywords: BioPhotonics; diffused reflectance; near-infrared spectroscopy; clinical diagnosis

1. Introduction

The human body is composed of different biomolecules. Broadly, they are classified into four large groups, carbohydrates, fats, proteins, and nucleic acids, with minor concentrations of other signaling molecules such as hormones, etc. Each of the biomolecules in turn comprise of four key elements—Oxygen, Carbon, Hydrogen, and Nitrogen. The composition of biomolecules that make up different tissues varies both between the tissue types spatially and varies within the same tissue temporally. This spatial and temporal variation is essential for healthy functioning of the human body. For example, blood has a different composition to muscles which is different from bones and is spatially segregated. Even within a tissue, there are spatial variances at different scales. A good example of such spatial variation in composition is a skeletal muscle that is attached to bones via tendons. The muscle itself is made of sarcomeres and myofibrils which differ in their molecular composition. Tissues develop variations in their biochemical composition over time like the healing of an open wound where a clot develops by the development of fibrin. White blood cells and macrophages clear the damaged tissue and foreign bodies. Over the healing process, new capillaries grow with new fibrous tissue development and skin recovery. The site of the biological process is not altered, but the composition changes temporally. During a diseased state, the composition of some of the biomolecules departs from the levels seen in healthy physiology. This shift from normal is the basis of clinical diagnoses such as blood tests, imaging techniques, etc. Clinical diagnostics are frequently invasive, cumbersome, and non-continuous. There is a need for repeated, reliable, non-invasive assessments of diseases that detects changes spatially like cancers and tumors, or temporally such as diabetes, sepsis, etc.

BioPhotonics diagnostic tools are capable of both spatial and temporal assessment of tissue, thus delivering the unmet need of repeated, reliable, non-invasive assessment in clinics. Light–tissue interaction is the basis of realizing diagnostic tools using BioPhotonics that are clinically relevant. When the biomolecules interact with light, the light is – absorbed, reflected, transmitted through, scattered—both linearly and non-linearly or transformed into a different colour through fluorescence by the biomolecule. Further, there can be

Citation: Krishnamoorthy, S. Light for Life—Optical Spectroscopy in Clinical Settings. *Eng. Proc.* **2023**, *34*, 27. https://doi.org/10.3390/HMAM2-14274

Academic Editor: Vijayakumar Anand

Published: 21 March 2023

heating or vibration-producing acoustic effects in the biomolecules [1]. Each of these light-tissue interactions is utilized in BioPhotonics to develop an effective diagnostic tool tailored to the clinical problem at hand. In this paper, we will focus on optical spectroscopy, where a unique optical spectrum of the biomolecule of interest is extracted using light. This optical spectrum is then used to identify disease states. To understand the generation of the optical spectrum, we note that different wavelengths of light interact with the biomolecules differently, initiating either electronic transitions or exciting rotational or vibrational modes of the chemical bonds. Since each of the biomolecules is chemically distinct, they generate unique optical spectra. For obtaining the optical spectrum of the biomolecule of interest a suitable optical source is required. To select the appropriate source one has to be mindful of the safety limits in terms of the optical power. Also, one has to note that light of certain wavelengths, such as X-rays and far UV are harmful to the tissue due to their ability to ionize.

In this paper, we look at the role of optical spectroscopy in two scenarios as shown in Figure 1. First in oral cancer-where spatial classification of tissue health is critical. Secondly in fetal hypoxia assessment-where a temporal evolution of biomolecules is monitored. Both these cases are based on the projects in the Author's group (Biophotonics@Tyndall) where the Author is involved.

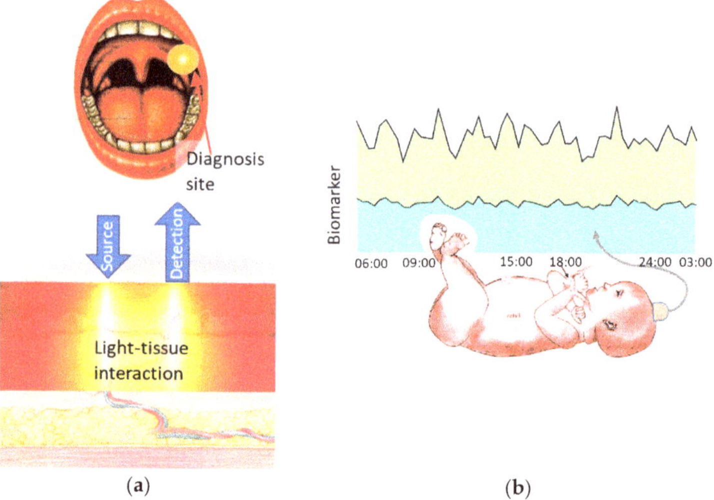

Figure 1. Biophotonics tools for diagnosis of diseases. (**a**) Spatial assessment of legions; (**b**) temporal monitoring of hypoxia in babies. Adapted with permission from Refs. [2–4] Copyright 2015 New England Journal of Medicine.

2. Spatial Assessment—Oral Cancer

Oral squamous cell carcinomas (OSCC) manifest in the early stages as painless white or red plaques in the mouth. Dysplastic cells under the oral skin accumulate many genetic alterations and mutations. This site now attracts macrophages while the immune response is suppressed. This site evolves into a high-grade dysplasia and then into invasive caner [5]. Metabolism in this site is different to the healthy tissue in other parts of the oral cavity. The detection of such sites by local swab sampling followed by histopathology is the current diagnostic gold standard. However, swab samples only access the superficial cell layers.

Optical spectroscopy can be designed to penetrate the tissue to a depth, non-invasively, where dysplasia might exist. Further, a multimodal approach that involves diffused reflectance spectroscopy (DRS), Raman spectroscopy, and fluorescence spectroscopy is used to target biomolecules accurately that are effective in the recognition of cancerous sites spatially [6].

3. Temporal Assessment—Fetal Hypoxia Assessment

Lack of oxygen, or hypoxia, in the baby during labour could result in death. The babies that survive this ordeal could be faced with developmental problems in their brain like cerebral palsy. Assessment of a hypoxic state is a critical clinical diagnostic event that allows the obstetrician to decide in favour of a Cesarean section (C-section). This time-critical diagnosis is currently done by monitoring fetal heart rate (FHR), which is an overall fetal well-being parameter, through fetal scalp electrodes or cardiotocographs (CTG). Oxygen content in the blood, SpO2, is a better correlate of hypoxia in the fetus and it is measured through fetal pulse oximetry. The current gold standard of hypoxia diagnosis is fetal blood sampling (FBS) where lactate and pH are measured in the fetal blood obtained from the fetal scalp using a scalpel. However, this method is invasive and can only be performed sporadically [7]. Thus, there is an unmet need for a non-invasive, continuous diagnosis of hypoxia where the changes in hypoxia related biomolecules, namely lactate concentration and pH, concentrations are monitored in real-time during labour. This will aid in the assessment of onset of hypoxia in babies.

For spectroscopic assessment of hypoxia, lactate and pH levels need to be determined. Light in the long wavelength near-infrared (LW-NIR) spectrum, i.e., 1350–2500 nm, is a well-suited spectrum for the detection of these biomolecules. LWNIR region consists of combination and secondary absorption bands of small molecules like lactate, water, glucose and lipids. Water forms 70–90% of all our tissues. Lactate spectral features are superimposed on large water absorption regions, so the detection of lactate demands high sensitivity in detection in this region. The spectral features are broad and have many interferences for other molecules like glucose etc. Thus, a good spectroscopic tool is required to assess hypoxia in this region [8,9].

4. Conclusions and Future Outlook

In this paper, a perspective of BioPhotonics is based on clinically relevant, non-invasive diagnostic tools. Specifically, diagnostic tools with the ability to find malignant lesions in Oral cancer diagnosis earlier and with a higher accuracy was discussed that focused on spatial detection. Next, a tool to diagnose fetal distress during labour was discussed, including the challenges of building such a BioPhotonics tool. When implanted, this would result in a safer delivery both for the mother and the infant. Both these cases have a transformative impact on clinical diagnostic practice.

Looking ahead, BioPhotonics-based diagnostic tools provide a non-invasive, sustainable, reliable way to assess diseases as they evolve over both space and time. They will provide non-destructive, real-time tissue diagnosis for both daily life as wearables and as tools during surgeries.

Funding: This research was funded by the SFI-15/RP/2828 SFI Professorship Award and EI-CF20211693B EI commercialization fund.

Institutional Review Board Statement: Not applicable.

Informed Consent Statement: Not applicable.

Data Availability Statement: No new data was created for this paper.

Acknowledgments: The author thanks the Biophotonics group@Tyndall National Institute and Irish Photonics Integration Center, South Infirmary Victoria University Hospital, Cork University Hospital, ENTO Research Institute, University College Cork, Dental School and Hospital, University College Cork, Cork Univ. Maternity Hospital (Ireland) and The INFANT Research Ctr. for providing the facilities and support for this work. Specifically, the author thanks the contributors Siddra Maryam, Marcelo Saito Nogueira, Rekha Gautam, Sanathana Konugolu Venkata Sekar, Kiang Wei Kho, Huihui Lu, Richeal Ni Riordain (ENTO Research Institute, University College Cork, Cork University Dental School and Hospital), Linda Feeley (ENTO Research Institute, University College Cork, Cork University Hospital), Patrick Sheahan (South Infirmary Victoria University Hospital, Cork University Hospital), Urbashi Basu (National Center for Biological Sciences, India, Princeton

University, USA), Francesca Di Croce (University of Pavia, Italy), Walter Messina, Cleitus Antony, Paul Townsend, Fergus P. McCarthy (Cork Univ. Maternity Hospital (Ireland), the INFANT Research Ctr., Univ. College Cork (Ireland), Ray Burke and Stefan Andersson-Engels.

Conflicts of Interest: The author declares no conflict of interest.

References

1. Tuchin, V.V. Tissue optics and photonics: Light-tissue interaction. *J. Biomed. Photonics Eng.* **2015**, *1*, 98–134. [CrossRef]
2. Tik, J. The Tonsils and Other Soft Tissue Structures of the Mouth. Doctor's Office Poster. Available online: https://www.flickr.com/photos/jantik/54322517/ (accessed on 14 February 2023).
3. Stankovic, K.M.; Tan, O.T.; Sadow, P.M. Case 36-2015: A 27-year-old woman with a lesion of the ear canal. *N. Engl. J. Med.* **2015**, *373*, 2070–2077. [CrossRef] [PubMed]
4. Fatema, N.; Acharya, Y.; Al Yaqoubi, H. Amniotic Band Syndrome: A Silent Knife In-Utero. *Nepal Med. Coll. J.* **2019**, *21*, 153–159. [CrossRef]
5. Rangel, R.; Pickering, C.R.; Sikora, A.G.; Spiotto, M.T. Genetic changes driving immunosuppressive microenvironments in oral premalignancy. *Front. Immunol* **2022**, *13*, 147. [CrossRef] [PubMed]
6. Maryam, S.; Nogueira, M.S.; Gautam, R.; Krishnamoorthy, S.; Venkata Sekar, S.K.; Kho, K.W.; Lu, H.; Ni Riordain, R.; Feeley, L.; Sheahan, P.; et al. Label-Free Optical Spectroscopy for Early Detection of Oral Cancer. *Diagnostic* **2022**, *12*, 2896. [CrossRef] [PubMed]
7. Al Wattar, B.H.; Honess, E.; Bunnewell, S.; Welton, N.J.; Quenby, S.; Khan, K.S.; Zamora, J.; Thangaratinam, S. Effectiveness of intrapartum fetal surveillance to improve maternal and neonatal outcomes: A systematic review and network meta-analysis. *CMAJ* **2021**, *193*, E468–E477. [CrossRef] [PubMed]
8. Krishnamoorthy, S.; Burke, R.; Mc Carthy, F.; Andersson-Engels, S. Beyond oxygen in-vivo long-wavelength near infra-red spectroscopy for hypoxia assessment. In Proceedings of the European Conference on Biomedical Optics, Munich, Germany, 20–24 June 2021.
9. Krishnamoorthy, S.; Basu, U.; Kho, K.; Gautam, R.; Di Croce, F.; Messina, W.; Antony, C.; Townsend, P.; McCarthy, F.P.; Andersson-Engels, S.; et al. Non-invasive continuous hypoxia assessment in intra-partum fetus through long wavelength near infrared spectroscopy. In *SPIE Proceedings, Proceedings of Photonic Instrumentation Engineering X, San Francisco, CA, USA, 30 January–1 February 2023*; SPIE: Bellingham, WA, USA; Volume 12428, pp. 176–178.

MDPI

St. Alban-Anlage 66

4052 Basel

Switzerland

www.mdpi.com

Engineering Proceedings Editorial Office

E-mail: engproc@mdpi.com

www.mdpi.com/journal/engproc